**MILITARY PROFESSIONALIZATION
AND POLITICAL POWER**

SAGE SERIES ON ARMED FORCES AND SOCIETY

INTER-UNIVERSITY SEMINAR ON ARMED FORCES AND SOCIETY

Morris Janowitz, *University of Chicago*
    Chairman and Series Editor

Charles C. Moskos, Jr., *Northwestern University*
    Associate Chairman and Series Editor

Sam C. Sarkesian, *Loyola University*
    Executive Secretary

Also in this series:

HANDBOOK OF MILITARY INSTITUTIONS
    Edited by Roger W. Little
MILITARY INSTITUTIONS AND THE SOCIOLOGY OF WAR:
A Review of the Literature with Annotated Bibliography
    by Kurt Lang

MILITARY PROFESSIONALIZATION AND POLITICAL POWER

BENGT ABRAHAMSSON

with a Foreword by Morris Janowitz

SAGE PUBLICATIONS
Beverly Hills / London

Copyright © 1972 by Inter-University Seminar on the Armed Forces and Society

All rights reserved. No part of this book may be reproduced or utilized in any form or by any means, electronic or mechanical, including photocopying, recording, or by any information storage and retrieval system, without permission in writing from the publisher.

For information address:

SAGE PUBLICATIONS, INC.
275 South Beverly Drive
Beverly Hills, California 90212

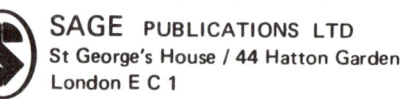

SAGE PUBLICATIONS LTD
St George's House / 44 Hatton Garden
London E C 1

Printed in the United States of America
International Standard Book Number 0-8039-0138-0
Library of Congress Catalog Card No. 79-172472
First Printing

CONTENTS

Acknowledgements 7
Foreword by Morris Janowitz 9
Introduction: The study of military professionalization 12

PART I: THE EMERGENCE AND HISTORICAL TRANSFORMATIONS OF THE MILITARY PROFESSION (PROFESSIONALIZATION$_1$)

One   Social change and the military profession: the emergence of technical and organizational expertise 21

Two   Social recruitment: increasing heterogeneity 40

PART II: HOMOGENIZATION OF OUTLOOKS AND BEHAVIOR (PROFESSIONALIZATION$_2$)

Three   Professional socialization: theory, ethics, and corporateness 59

Four   Occupational values and the military mind 71

Five   Nationalism: axiom of military ideology 80

Six   Images of human nature 85

Seven   The military theory of risks: alarmism 87

Eight   Authoritarianism 93

Nine   Some implications for military occupational prestige and intellectual recruitment 96

PART III: PROFESSIONALIZATION, INFLUENCE, AND POWER

Ten   Elements of military conservatism: traditional and modern   101

Eleven   Normative influence and the institutionalization of professional values 112

Twelve   The military profession and political power: resources and their mobilization 140

Thirteen Professionalization, power and civilian control: summary
and conclusions 151
References 164
Name index 173
Subject index 180

ACKNOWLEDGEMENTS

I owe my thanks to a number of friends and colleagues who have contributed helpful criticisms and constructive suggestions, as well as other assistance. Professor Morris Janowitz enabled me to stay one year at the University of Chicago as a Post-Doctoral Fellow while working out part of the manuscript. At the University of Stockholm, Professor Gunnar Boalt has taken a stimulating interest in my work.

Kerstin Bohm, Robert Erikson, Olof Fråndén, Jack Hammond, Walter Korpi, Jeremy Kunz, Moshe Schwarz, and Sune Sunesson all have read preliminary chapter drafts and provided fresh viewpoints which I hope they will not find too misused. Joy Sarkar and Tom Burns checked the English. Tom has also read the final manuscript and suggested a number of highly useful changes of content and style.

I have been able to participate in the work of the Research Committee on Armed Forces and Society of the International Sociological Association; its sessions in Evian, 1966, London, 1967, and Varna, 1970 gave many impulses for my work, as did my discussions with several of its members, among them Fabrizio de Benedetti, Philippe Schmitter, Jerzy J. Wiatr, and P. Zhilin. Although I have profited from the judgments of many competent persons in the field of military sociology, a number of schortcomings that are entirely my own most certainly remain.

Finally, Ulla, Malin, and Kalle have had to put up with my running away from a number of family functions. I admire their tolerance.

Stockholm, April 1971

Bengt Abrahamsson

FOREWORD

Bengt Abrahamsson is a member of the new generation of European sociologists who are thoroughly familiar with the historical and classical literature of sociology and who are well trained in the practice and evaluation of empirical research. His sense of theoretical, substantive, and policy relevance has led him to concentrate over the years on the study of the military profession and military organization, both in Sweden and more generally in industrialized nations. In *Military Professionalization and Political Power,* he presents a highly systematic analysis of the military which is grounded boldly in organizational theory and which synthesizes carefully the extensive body of historical and contemporary research materials. It is a book which deserves and will command widespread attention.

He makes use of the generally accepted sociological format for the analysis of a professional group in a bureaucratic setting as he probes deeply into a number of leading hypotheses which are to be found in the existing literature on military institutions. The study starts with an analysis of the transformation of the technology of war and the resulting consequences on the structure of military organization and the goals of the military profession. He proceeds to examine the changing patterns of social recruitment and professional socialization of officer cadres. The object of his analysis is to describe and account for the ideology — both professional and political — of the military officer, and to assess their political power position in multi-party states. Abrahamsson, while fully aware of national differences, presses for the broadest conclusions about the military profession within his focus which is mainly on Western Europe and the United States.

A crucial aspect of the study centers on a set of hypotheses which hold that longer length of service and higher position in the military hierarchy produce (a) a stronger sense of military alarmism (increased expectation in the outbreak of war) and (b) a more marked sense of political conservatism. In short, these are the attitudes and beliefs which military professionalism does in fact produce. Of course, Abrahamsson is aware that these consequences are the outgrowth both of self selection in entrance and promotion in the military as well as the positive impact of career experiences and indoctrination. Moreover, ideological orientations need to be seen in the light of the intellectual capacities of officer personnel. Drawing on the documentation from Swedish sources — but presumably comparable for other industrialized nations, he is able to conclude that the intellectual and academic level of the military officers is lower than for those men entering other professional groups. Such a frame of reference supplies the context in which the author assesses the strong power position of the military profession in the contemporary setting.

As a result, Abrahamsson is explicit in this rejection of Samuel Huntington's policy orientation that professionalism insures civil-control. First, for Abrahamsson, Huntington is empirically incorrect; the impact of professionalism, if left unchecked, runs the risk of separating the military from the socio-economic balance of civilian society. Second, Huntington defines the essential problem out of existence; if the military break, thwart, or pervert civilian control, they are defined as being unprofessional.

Thus, civil control must rely both on objective control and subjective control. Both Abrahamsson and I find it appropriate to define objective control and subjective control in a fashion at variance with the usage by Samuel Huntington. Objective control implies legislative and administrative institutions and a political base for making them effective. Subjective control implies professional norms and values which reflect the education and career system of the military profession and impact of contact with civilian society. In short, the author presents a viewpoint concerning the military profession which is currently being argued for many other professional groups; namely, external institutional civilian and social control (objective control) is only partial and insufficient. Civilian control must operate to develop subjective control, namely, a set of values and norms which are compatible with the social and political decision-making process of the larger society. Professionalism is not merely concerned with procedures but with societal goals and priorities. In the case of the military profession, changed technology and new values have transformed international relations. The result is an overriding necessity to develop new forms for resolving international conflict. These requirements are yet to be adequately incorporated into the military as a professional group; and they must be incorporated if it is to be a professional group.

Such a sociological interpretation of the military has central policy implications since the trend in Western Democracies is toward smaller long service and fully professional armed forces; the trend is toward what is called a "volunteer" armed force. Manpower and organizational requirements of modern weapons systems and the new logic of international relations propel this transformation. In the end, the implication of Abrahamsson's analysis converges with the concerns of many self critical and alert professional officers, namely that the emerging trends in military organization carry the likelihood of fundamental separation of the military from civilian society, both institutional, ideological, and political.

The volume will be noted for its clarity of formulation, precision of conception, and careful exposition. It reflects the Swedish intellectual posture in social research in its detachment and objectivity and its emphasis on an international approach. It is representative of the increasing efforts of sociologists to study the military in a comparative perspective, and as scholars who seek to be members of an international community — or who at least are keenly aware of their national attachments and allegiances. Abrahamsson has been an active participant in developing the research activities of the Research Committee on Armed Forces and Society of the International Sociological Association, and this book is a splendid example of the type of research this group seeks to stimulate and assist.

Morris Janowitz, Chairman
Research Committee on Armed Forces and Society
International Sociological Association

University of Chicago
Chicago, Illinois

INTRODUCTION:
THE STUDY OF MILITARY PROFESSIONALIZATION

According to one estimate, the world total of defense expenditures at the mid–1960's was in the order of 118 billion US dollars, a sum roughly equal to the gross national products of India, Japan, Poland, and Sweden added together.[1] The direct, day-to-day control of the major part of these resources is carried out by professional military men.

Thus, in addition to its potential coercive influence, the military profession commands considerable economic power. Whether deliberately intervening or not, the military profession is in most countries today one of the most important political pressure groups by virtue of the huge resources vested in it.

This is a book on the military profession and the political implications of military professionalization. Although I will make a few occasional digressions to countries of the Third World, the emphasis will be on the industrialized countries of the North American and European areas. In these regions, and especially in countries with a highly developed capitalist economy (such as the United States and Sweden), the military profession has emerged in its most pure form, as a group of technically and organizationally trained experts in the management of violence, held together by the bonds of common education, corporate practice, and professional ethics.

The military profession today differs in many important respects from the military of the late eighteenth century. It is recruited on the basis of education and skill rather than on the basis of social origins. Its corporateness stems from the common educational and intra-professional experiences, and not from similarity of social class (as we shall see, contemporary military recruitment is highly heterogeneous). Military men work on a full-time basis, instead of regarding military service as a part-time vocation or hobby, as did their eighteenth-century colleagues.

Modern armies permit mobility between ranks, and occasionally even enlisted men may advance to high military posts (usually after having received additional education); in eighteenth century continental armies, even wealthy bourgeois officers – whenever they managed to make their way into the military – would frequently find their careers blocked because of the caste prerogative of the nobility. Today, captains and

[1] Emile Benoit and Harold Lubell, "The World Burden of National Defense", in Emile Benoit (ed.), *Disarmament and World Economic Interdependence,* Oslo, 1967: Universitetsforlaget, tables 2 and 4.

colonels reach their ranks on professional merits, without paying for them; in the pre-revolutionary French army, a captaincy could cost between 6.000 and 14.000 livres, and a colonelcy (giving command over an infantry or a cavalry regiment) between 25.000 and 120.000 livres (for "cheap" infantry and "expensive" cavalry regiments, respectively).[2]

Seen in a historical perspective, military professionalization has been accompanied by a growing overlap between strategy and politics until today, with the prospects of nuclear confrontation, even local military decisions may have large-scale political repercussions. As a result, modern military officers find their roles encompassing a number of political aspects and, in fact, often receive training with the purpose of making them able to deal with highly complex politico-strategic issues.[3] Thus, instead of segregating officers from politics, military professionalization has brought about a fusion of military and political roles.

In social science literature, one frequently finds references to the alleged "affective neutrality" of professional groups, i.e., the notion that the primary function of professionals is impartial and neutral service in the interest of the public.[4] This assumption is to a great extent invalid, because the process of professionalization tends to implant not only public-service oriented values, but also loyalty and devotion to the profession itself and the values and goals it represents. Since these goals and values may at times be contrary to the interests of major public groups (and occasionally of the population at large), it follows that professionalization may result in behavior contrary to the norms of impartial and neutral service.

To exemplify, we hardly expect the major values of the military profession to be disarmament, internationalism, and trust in the good-will of other nations, but rather their opposites. If this assumption is correct — and I will support it with more detailed arguments and data later on — then a number of problems will arise in civil-military interaction when civilian groups advocate security strategies based on international cooperation, or when such groups seek to restructure the economy by reducing military spending (in order to increase non-military industrial production, social security and educational programs etc). Thus, whereas military professional values are functional to the carrying out of military tasks, they may be highly dysfunctional to policies aimed at increasing social equality or at unorthodox security arrangements. To view the military profession only from the "neutral expert" perspective is to perpetuate a number of invalid assumptions. Professionalization creates experts, but it hardly guarantees their neutrality, particularly not in areas where they themselves are most active.

1. Defining "profession": the typological vs. the gradualistic approach

Sociological writers in the field of professionalization may be distinguished according to the approach chosen in defining the key concept of

[2] Samuel F. Scott, *The French Revolution and the Professionalization of the French Officer Corps, 1789–1793*. Paper, 7th World Congress of Sociology, Varna, Bulgaria, 1970, p. 7. Scott reports the yearly income of provincial nobles to have been around 10.000 – 15.000 livres, and for "minor nobles of the sword" (to be distinguished from the "noblesse de robe") between 600 and 1200 livres. This latter group, being most numerous, was obviously blocked from quick careers because of unsufficient financial resources, and had to make their limited advancement "on the strength of their own merits" (*ibid.*, p. 13).

[3] See John W. Masland & Laurence I. Radway, *Soldiers and Scholars*, Princeton, 1957: Princeton University Press, ch. 1; also ch. 11 below, and Appendix.

[4] See for example R. D. McKinlay, *Professionalization, Politicization and Civil-Military Relations*. Paper, The Perceived Role of the Military, Social Science Symposium, Bandol, France, 1970, p. 6; and Suzanne Keller, *Beyond the Ruling Class*, New York, 1968: Random House, p. 4.

"profession" (to the extent that they at all make their definitions explicit). The first approach, partly in the tradition of Carr-Saunders[5], may be termed qualitative or *typological*. It tries to identify major characteristics separating occupational classes from each other, and strives to find criteria (sufficiently clear for analytical purposes) that separate the class of "professions" from that of "non-professions". Some writers have carried the typological approach still further, discriminating also between various sub-classes within the general class of professions. For example, Albert J. Reiss, Jr. has provided the following list of "sub-professions":

[5] A. M. Carr-Saunders & P. A. Wilson, *The Professions,* Oxford, 1933: Clarendon Press.

1. Established professions (medicine, law, the clergy)
2. New professions (e.g., chemists, engineers)
3. Semi-professions (with relatively little theoretical basis, e.g., pharmacists, opticians, social workers)
4. Would-be professions (aspiring to professional status, e.g. public relations men, undertakers)
5. Marginal professions (relatively skilled personnel working under the supervision of a professional, e.g., laboratory technicians, psychological test personnel, draftsmen).[6]

The fruitfulness of the typological approach hinges on the possibility of finding characteristics that clearly delimit one class of occupations from another. If the classificatory criteria chosen are not exclusive of one another, the classification is bound to be confused, and one may raise doubts as to the empirical utility of such a scheme. Such doubts seem to be justified when considering Reiss's list of "sub-professions". For example, it is hardly apparent why engineering is classified as a "new" profession, whereas social work is not, or why undertakers are assumed to aspire to professional status whereas opticians are not.

[6] Albert J. Reiss, Jr., "Occupational Mobility of Professional Workers", *American Sociological Review,* vol. 20, 1955, pp. 693–700.

A second and better approach, which may be called quantitative or *gradualistic,* conceives of all occupations as being more or less professionalized.[7] What is a profession, and what is not, is determined by cutting the professionalization continuum at some point, labelling occupations above this point "professions". The professionalization dimension is by this approach taken to represent a small number of universally applicable variables (for example, degree of theoretical basis, ethical rules, and corporateness; see below), felt by the theoretician to be important for the analysis of occupations.

The typological approach is deterministic, always assigning the value 1 of a certain variable to a certain occupation; its major effort is directed towards finding a number of such variables permitting the grouping of occupations into "families" and "species". The gradualistic approach is stochastic, selecting a small number of universal variables, and then assigning values between 1 and 0 of those variables to the various occupations. Using the gradualistic approach (employing the three variables indicated above), we would find at the lower end of the professionalization continuum a number of low-skilled occupational groups such as agricultural workers, window cleaners, bus drivers, and hot dog vendors, and at

[7] See Ernest Greenwood, "Attributes of a Profession", in S. Nosow & W.H. Form (eds.), *Man, Work, and Society,* New York, 1962: Basic Books; and H.M. Vollmer & D.L. Mills (eds.), *Professionalization,* Englewood Cliffs, New Jersey, 1966: Prentice-Hall.

the upper end occupations such as medicine, law, the clergy, and the military.

Some merits of the gradualistic approach are, (1) that it makes professionalization independent of the aspirations (or other "private" variables) of the members of the profession (as implied in the term "would-be profession"), thus allowing classification without the use of subjective data; (2) that the classification is independent of authority relationships within the profession (implied in "marginal profession"): we may want to establish empirically the degree of independence of professional members and should not make this a matter of definition; and, (3) that the classification can be made independently of the attitudes to the occupation among the public (partly implied in the term "established profession").

2. Some major terms

Following the gradualistic strategy, the present work means by a profession an occupation whose members (a) possess a high degree of *specialized, theoretical knowledge,* plus certain methods and devices for the application of this knowledge in their daily practice, (b) are expected to carry out their tasks with due attention to certain *ethical rules* and, (c) are held together by a high degree of *corporateness* stemming from the common training and collective attachment to certain doctrines and methods.

When I speak about "the military profession", I mean the most professionalized part of the military occupational group, i.e., the officer corps. The reader should be aware that, as suggested above, this represents a relatively arbitrary cutting-point on the military occupational spectrum. Since we may distinguish between *more or less* professionalized military men according to the degree to which they hold the characteristics (a), (b), and (c) above, the discussion to follow will be more or less applicable to other strata than the officer corps, according to their level of professionalization. For instance, the particular set of values and outlooks described as "the military mind" may be assumed to be most prevalent among commissioned officers, somewhat less among non-commissioned officers, and still less among enlisted men. Common to all three groups, however is the *relationship* between the process of professionalization and the holding of indigenuous values and outlooks.

Terms such as "the military" or "the military establishment" are wider in scope than "the military profession" and, when used in the text, include officers, NCO's and soldiers of lower ranks as well as a certain number of civilians in semi-military positions. (Most military establishments use specially employed civilian personnel for various technical and auxiliary tasks, usually on relatively low levels, such as telecommunication technicians, secretaries, and staff assistants). These concepts do not refer to any members of the civilian political leadership, such as secretaries of defense, or other civilian agents exercising — or supposed to exercise — control over the military.

The concept of *professionalization* has been employed in the literature usually with two different meanings, both of which will be of importance for the present discussion.

In its most common usage, professionalization is equivalent with *professional socialization,* that is, the process by which *individuals* are being transformed from a state of relative unawareness of the theoretical and practical problems of the profession's issue area, to the state of acute awareness of such problems. Part of this socialization may take place before induction (in which case we speak about anticipatory socialization); part of it will be an effect of education, training, and interaction with colleagues during the professional member's daily activities (intra-professional socialization). In this sense, professionalization refers to the individual, micro-level changes in outlooks and behavior among the members of the profession. Some questions relating to this will be treated in chs. 3–9.

In a somewhat less common usage, professionalization refers to the *historical transformations* of a particular occupational group, under the impact of major political, economical, and technological developments. As already mentioned above, during the last two hundred years a number of significant changes have occurred within the military, transforming the officer corps from a group of part-time employed, ascriptively recruited soldiers, to a well-educated, technically and managerially trained corps of experts, recruited on the basis of achievement and skill. These transformations are reflections of major social changes, such as the emergence of nation-states, the industrial revolution, the decline of the nobility, and various technological developments (such as new weapons, innovations in communication and transportation); they will be discussed in more detail in Part I of this book (chs. 1 and 2). Similar analyses, using historical sources, could be made of the developments of other professional groups (such as the professions of medicine, clergy, law, engineering, and journalism).

It will be convenient for the following discussion to label the historical transformations of the military profession *professionalization*$_1$ (or p_1 for short), and the process of transformation of individuals *professionalization*$_2$ (p_2). The characteristics of a profession at a given time will be a function of both processes. Adaptations on the macrolevel (p_1) will have consequences for micro-level socialization (p_2). Thus the outlooks and behaviors of the members will be a reflection of the social forces affecting the profession. We arrive at the composite picture of connections between major social forces, professionalization$_1$, professionalization$_2$, and feedback processes as exhibited by diagram I.

DIAGRAM I

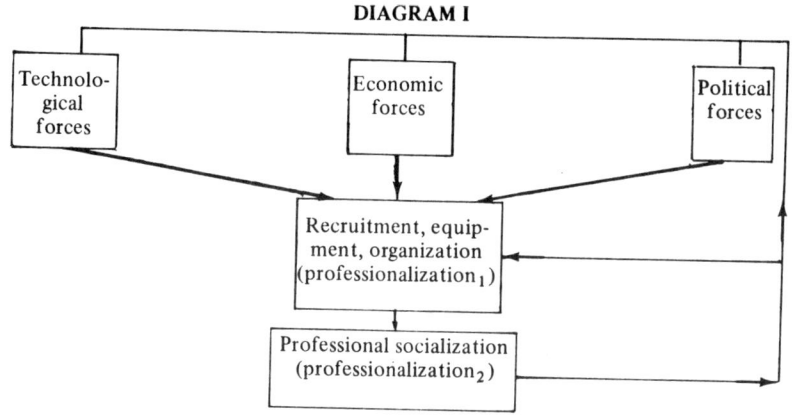

However, one characteristic of a professional group is that it does more than simply adjust to such forces: it also tries to anticipate, influence, and regulate them in order that they be as consistent as possible with the profession's corporate interests. Because of their expert authority, high social prestige, and social connections, professions are more able than most occupational groups to exert influence on other social elites and to affect political, economic, and technological forces in a society.

My general thesis is simple. Military professionalization$_1$ involves the creation and maintenance of a complex, effective, and well-organized social institution. Military professionalization$_2$ involves the indoctrination and internalization of certain values, outlooks, and behavior elements. Professionalization$_1$ has created a politically powerful and often highly independent social structure; and professionalization$_2$ molds the individuals who are going to man it. *To the extent, then, that military professionalization in the above senses is effective* – and to the extent that there are differences between military and civilian values and objectives – *civilian control of the military establishment will be impaired*. In other words, in a situation where civilian authorities want to pursue policies that are in disagreement with those preferred by the military, those authorities will meet greater resistance from a high-professionalized than from a low-professionalized officer corps. Or again: military men are not and cannot be neutral and objective servants of the state: they hold certain beliefs, have certain corporate interests and can be expected to favor and to pursue political actions that are consistent with those beliefs and interests.

3. Military professionalization and political intervention

The military is an instrument of the civil state for the exercise of violence, supposedly against other states. In theory at least, control of domestic violence is assumed to be the responsibility of the police, militia, and other para-military forces, while the military is trusted with defending the state against external aggression. This is the essence of the idea of civilmilitary relations in modern countries and the official raison d'etre for the military establishment.

The military is expected to be subordinated to civil authority. Although this relationship has always been problematical, the problems seem to have grown considerably in scope during the present century. In no small measure, this is due to the impact of military professionalization.

With the acceleration of military technology, weapons of mass destruction, and the complexity of military organization, the power potentials of the military profession have increased. The borderline between strategy and politics, which has never been completely clear, has become blurred to an extent heretofore unknown. Although this trend is obvious in the military establishments of the major world powers[8], the political role-expansion of the military officer appears to be a phenomenon of major importance in most countries (since changes in weapons technology and military organization in the United States and the Soviet Union are bound to have repercussions on a wider scale).

[8] See John W. Masland & Laurence I. Radway, *Soldiers and Scholars,* ch. 1.

Military coups d'etat have been rare in the European and North American zones. But the military profession is able to exert political power in many other and more subtle ways than through armed intervention: through political bargaining, trading military installations and investments for support from politicians; through carefully researched and eloquently presented policy recommendations, supported by the weight of military professional expertise; and through lobbying and public relations work. This is facilitated by the fact that many politicians remain indifferent to military problems; and this, in turn, may be caused by the fear of appearing unpatriotic if they question military proposals, or by the lack of the necessary military knowledge, or by the fact that military issues often do not "pay" as instruments for attracting votes (since the general public is frequently unresponsive to foreign policy and security arrangements, paying greater attention to issues like taxation and education).

It is natural that professionals be committed to what they consider to be the "right" utilization of the resources of the profession, to use such resources in ways approved of by the profession itself. Opposition to professional solutions are likely to be regarded by the members as illegitimate attempts at limiting professional independence.

However, the professional independence of the military acquires special significance because of the size of military economy, the scope of the military organization, and the annihilation that will follow if the full resources of the military establishments come into use. I believe that a first-rate social problem in the area of civil-military relations is created by the fact that the full effects of military professionalization for civilian control have not been adequately analyzed.

4. The plan of this book

To explore these effects, and the conditions leading to them, the following areas are important.

I. Professionalization$_1$

1. The *historical processes* leading to the emergence of the military profession as a corps of specialized experts in the management of violence (ch. 1)

2. The *changes in social recruitment* of the military profession. Traditionally, much interest has been centered on the social origins of the military, under the assumption that such information would facilitate the explanation and prediction of military political behavior. Today however, it seems more reasonable to expect the impact of professionalization to be a stronger determinant of such behavior. This issue will be discussed at some length. In this connection I will also draw together and comment on some fairly recent data on the social origins of the military in a number of countries (ch. 2).

II. Professionalization$_2$

3. Several aspects of professional life may be assumed to influence the values, outlooks, and behavior of the profession's members. I have chosen to discuss these aspects in fairly general terms, in the hope that the discussion will be applicable to other fields than the military; consequently, I have been rather restrictive in the use of military examples. (ch. 3).

4. There has been much discussion and some controversy on the "military mind". In ch. 4, I will attempt to assess the importance of that concept, to define it, and to indicate some of its major dimensions. A number of short chapters will then be devoted to each one of these components, i.e., nationalism (ch. 5), pessimistic beliefs on the nature of man (ch. 6), alarmism (ch. 7), and authoritarianism (ch. 8). Political conservatism is also part of this cluster, but will be left for special consideration in Part III (see below). In ch. 9, I will draw some implications of the military set of values for factors such as occupational prestige and intellectual recruitment.

In most of these areas, empirical data are too scarce to allow for more than a preliminary overview. I feel, however, that such an overview may be of use for directing attention to some important areas of research and, perhaps, in generating more empirical investigations than have been possible in the present study.

III. Professionalization and politics

Political conservatism is consonant with the dimensions of the military mind as described above. The hypothesis of military conservatism is supported by empirical data from a number of countries. These data are presented together with some attempts at explanation (ch. 10), forming the first of three chapters on the military profession and politics. I have found it convenient for theoretical reasons to distinguish between the military profession's *normative influence* and *political power,* the first concept representing the ability of the professional group to exert influence over social values (ch. 11), the second one referring to the profession's overcoming of resistance in actual decisionmaking concerning military objectives (ch. 12).

The book concludes with a summary chapter relating the four main concepts of professionalization$_1$, professionalization$_2$, normative influence, and political power to each other (ch. 13).

5. Limitations of the present study

In general, my aim has been to identify a number of *key variables* that I regard as important for the sociological analysis of the military profession. Wherever possible, I have tried to indicate the relationships between those variables and to substantiate the discussion by presenting empirical data. It must be emphasized, however, that the military profession is not exactly the easiest group to approach with questionnaires and interviews; the difficulties are especially great if the questions concern attitudes to politics and civil-military relations, whereas inquiries on less ego-involving items (e.g., social background) meet less resistance. The reader should not, therefore, expect too much from the evidence cited to support the theses advanced in the book. I have found it more important trying to present a fairly coherent theoretical discussion, supplying it with data wherever accessible, than to limit the discussion only to sections where first-rate data are available. I can only hope that the reader will share my view.

A few words of what this book is *not* about seem in order. This is not a book on the military-industrial complex, although the reader may find occasional references to that problem area.[9] Neither is it a survey of the various instances of military intervention in politics; such accounts may be found elsewhere for a large number of countries.[10] The focus on professionalization leads to an emphasis on what is common to military men regardless of their national identity and, consequently, brings forward the similarities at the cost of international differences which sometimes may be important. I believe, however, that the sociological analysis of the military can profit from a consideration of the "international invariances" of the profession as a general basis from which the deviations may later be worked out in greater detail.

[9] For a number of excellent discussions of the MIC, see *Report from Iron Mountain* (Swedish edition, *Angående möjligheten och önskvärdheten av fred,* Stockholm, 1968: Raben och Sjögren); John Kenneth Galbraith, *The New Industrial State,* Boston, 1967: Houghton Mifflin Co., esp. ch.xxix; and *ibid., How to Control the Military* (Swedish edition *Att hålla Pentagon i schack,* Stockholm, 1970: Tema). See also the articles on "American Militarism", *Look,* 1969, vol. 33, Nos. 16 and 17 (August 12 and August 26).

[10] For bibliographies, see Moshe Lissak, "Selected literature on revolutions and coups d'etat in the developing nations", in Janowitz, M. (ed.), *The New Military,* New York, 1964: Russell Sage Foundation; and Kurt Lang, *Sociology of the Military,* prepared for the Inter-University Seminar on Armed Forces and Society, 1969, esp. pp. 57–74.

PART I

THE EMERGENCE AND HISTORICAL TRANSFORMATIONS OF THE MILITARY PROFESSION

(PROFESSIONALIZATION$_1$)

ONE

SOCIAL CHANGE AND THE MILITARY PROFESSION: THE EMERGENCE OF TECHNICAL AND ORGANIZATIONAL EXPERTISE

Introduction

Professionalization is part of the general process of social differentiation and occupational diversification which, in turn, is a product of economic and technological development. Professionals form the "staff" of industrialized society, occupying roles as administrators and experts in areas such as education, science, engineering, and medicine. This chapter will cover the historical development of the military profession as the "staff" of the armed establishment, the instrument of territorial defense and expansion.

The emergence of an organized corps of officers constantly preoccupied with the preparation of war (rather than doing "emergency" duty in times of war, or having officership as a hobby) is an outcome of two wellknown historical processes. First, the centralization of state authority; second, the industrial revolution.

As exemplified by the European monarchies and revolutionary France during the eighteenth and nineteenth centuries, social differentiation and the development of commerce and industry was facilitated and accelerated by policies of national unification, removing obstacles such as regional trade tariffs, regional coinage, and settlement laws restricting the mobility of the labor force.[1] Sectional interests associated with the old feudal society became neutralized by the impact of centralized and bureaucratized power, serving the enterprises of the rising bourgeoisie.

The emergence of the centralized nation-state provided a primary raison d'etre for the standing army, recruited nationally rather than internationally and officered by the nobility, "driven by hunger from their cold castles"[2] and seeking posts that enabled them to stick to their traditions of chivalry and feudal honor. The military profession provided a suitable living, for a long time sheltered from middle class competition and conveniently free from educational demands. Noble origins were enough to qualify for officer positions in most European countries well into the nineteenth century.[2b]

The military profession became more complicated, however. Noble birth as a recruitment criterion slowly became obsolete as logistical

[1] For a discussion of this with particular reference to German unification, see Frederick Engels, *The Role of Force in History*, London 1968: Lawrence and Wishart, esp. pp. 66–69.

[2] Alfred Vagts, *A History of Militarism*, London, 1959: Hollis and Carter, p. 52.

[2b] The first and most dramatic break with this tradition came with the restructuring of the French officer corps during the Revolution. See Samuel F. Scott, *The French Revolution and the Professionalization of the French Officer Corps,* paper, Research Committee on Armed Forces and Society, Seventh World Congress of Sociology, Varna, Bulgaria, 1970.

development, mass-produced new weapons, and the rapid growth of the armies forced new requirements on formal military education and organizational training. Toward mid-1700, the hegemony of the gentry and nobility in the military establishments of Europe was meeting an increasing challenge through sons of bourgeois families, entering the artillery and engineering branches. The nobles were reluctantly forced to accept military education, and to join the military academies that were established in many countries around the turn of the eighteenth century.[3]

The military profession's emphasis on technical and organizational expertise is essentially a product of the early 1800's while other elements, such as the code of honor, can be traced much further back. Born in the military schools of France, England, and Prussia, military professionalization was nurtured by, as well as nurturing, early industrialization. Exported to the United States, the professionalization of the military in the new nation was marked by the foundation of the military academy at West Point in 1802. Almost a century later it made its appearance on the South American scene, here as in North America strongly influenced by Prussian and French military thought and developing in schools with an overwhelmingly technical orientation.

The centralized state, equipped with facilities for propagating and indoctrinating its goals and for insuring identification with those goals, laid effective claims on the participation of its citizens in defense or armed conquest. The nation-states commanded organizational resources to call up huge numbers of men, if necessary by coercion, for its mass armies. But this was not quite sufficient; they were required to wait for the means of arming and equipping those armies.

The means were provided by the industrial revolution. The development in machine productions created the basis for turning out the huge amounts of weapons and other equipment without which Napoleon's armies could not have been turned into such formidable instruments of conquest and terror.

Prior to the nineteenth century, wars had usually been expensive enterprises, and armies had been small. At Breitenfeld, Gustavus Adolphus commanded 39.000 men (reduced, however, by 16.000 already at the beginning of the battle through the desertion of Saxonians), against the forces of the German Catholic League of 32.000[4]. At the height of her armament efforts under Frederick William I (around 1715), Prussia expanded her army from 40.000 to 83.000 men. However, this was made possible only by subjecting the economy to great strain by spending about four or five times as much annually on the army as on all other state obligations.[5]

The combined effects of mass conscription (e.g., the law of "levée en masse" of 1793 in France) and the industrial production of weapons made warfare considerably cheaper and facilitated the creation of far larger armies for a given sum of money. At the same time, the rising class of capitalists and industrialists achieved an extra stimulus through the

[3] As late as 1864, Edwin von Manteuffel, Chief of the Prussian Military Cabinet, expressed as his opinion that it was unwise 'to require erudition from all officers', since 'the great majority . . . will consist always of qualified front officers and, for them, scholarly training is not necessary in such a high degree'. Quoted in Craig, G., *The Politics of the Prussian Army, 1640–1945,* Oxford, 1955: Clarendon Press, p. 234. Craig points out that "in reality, /Manteuffel's/ objection stemmed from the fact that the tightening of educational requirements represented a threat to the social cohesiveness, and hence the political reliability, of the officer corps". (*Ibid.*, loc.cit.)

[4] Jacques Boudet et al (eds.), *Arméernas Världshistoria,* part II Stockholm 1967: AB Svensk Litteratur, p. 204

[5] Craig, *op.cit.*, p. 8

mass manufacture of, and trade in, arms.[5b]

New requirements were put on the military. Warfare, which earlier had been something of a hobby or, at most, a part-time vocation for the nobility, became more complicated through the introduction of large numbers of lighter and more effective weapons; the French accomplishment at Valmy and the success of the highly mobile Napoleonic armies showed, for instance, the military gains that could be reached by the lighter field artillery. The size of the mass armies forced the development of organizational differentiation and subdivision. This, in turn, made new demands on careful pre-planning of operations, on liaison, and on the coordination of the various units and functions. The German General Staff in mid-1800 proved to be an effective solution to these new problems. The bureaucratization of military leadership, together with the appearance of steam-powered transportation and the electric telegraph, set the stage for a new type of officer — the military manager.

Today, the enormous expansion of the logistic functions of armed forces has made the military establishment almost a replica of civilian society. Further, the invention of the aeroplane and, later, the ballistic missile created a new concept — total warfare. Total warfare brought total defense; this, in turn, meant closer integration between the military and civilian sectors.

Within a period of less than two centuries, military amateurs have developed into military professionals. In most countries, the military profession can command greater economic resources than any other single vocational group; in a large number of states at the present time, the military's share of political power is at par with its share of economic resources, whether exercised directly as in many Latin American republics, in Greece, in the Middle East, and in a large number of the post-colonial nations, or indirectly as in the Eastern and Western industrialized states.

In the following pages, the historical trends sketched above will be developed somewhat more extensively.

1. Centralized state power

The emergence of the centralized state is usually exemplified by the reform policy of the French revolution. However, centralization, secularization, and bureaucratization had made their appearance far earlier. In France, Louise XIV laid the cornerstone for the authoritarian organization and unification which was later to be implemented with such effectiveness by Napoleon Bonaparte.[6] In Brandenburg Frederick William, the Great Elector, had come to power in 1640; he incorporated the German territories into one administrative unit, reorganized the financial system, and neutralized the opposition of the Estates of Cleves, Mark, and East Prussia.[7] His grandson, Frederick William I (1688-1740) founded the tightly knit Prussian bureaucracy and organized the military

[5b] As Hobsbawm has pointed out, "the iron and steel industries ... enjoyed no possibilities of rapid expansion comparable to the cotton textiles, and therefore relied for their stimulus on government and war". E.J. Hobsbawm, *The Age of Revolution: Europe 1789–1848*, London 1962: Weidenfeld and Nicolson, p.96

[6] See e.g. F. Ponteil, *Napoleon 1er et l'organisation autoritaire de la France*, Paris, 1956: Librairie Armand Colin

[7] Craig, *op. cit.*, p. 5

system later to be used by Frederick the Great against Austria and in the Seven Years' War.

The process of centralization gradually reduced the influence of local power centers. A sense of national identity began to replace the earlier, separatistic attachments to religious and provincial interests. In the words of Robert A. Nisbet:

> In the Middle Ages men were conscious of themselves as churchmen, guildsmen, members of this or that family or province, but never as a nation, much less as a people, with independent, corporate existence. Conceptualization of the people as an entity is a gradual process in modern history; historically its basis is, first, the atomization of the medieval *social* identities of individuals and, second, the centralization and nationalization of political power, thus creating a legal atmosphere within which socially detached masses of individuals could live and have identity.[8]

[8] Robert C Nisbet, *The Sociological Tradition*, New York, 1966: Basic Books, p. 121

The process of centralization was accompanied by the development of commerce, the improvement of printing, and urbanization. Improvement and extension of the infrastructure — roads, canals, ports — also contributed to weakening local and regional power concentrations. Marx expressed his admiration of the French revolution which like a "gigantic broom" swept away the localism of traditional France.

> The centralized state power with its ubiquitous organs of standing army, police, bureaucracy, clergy, and judicature... originates from the days of absolute monarchy, serving nascent middle-class society as a mighty weapon in its struggles against feudalism. Still its development remained clogged by all manner of medieval rubbish, seignorial rights, local privileges, municipal and guild monopolies and provincial constitutions. The gigantic broom of the French Revolution swept away all these relics by bygone times, thus clearing simultaneously the social soil of its last hindrances to the superstructure of the modern state edifice raised under the First Empire...[9]

[9] Lewis S. Feuer, ed., *Marx and Engels: Basic Writings on Politics and Philosophy*, Garden City, 1959: Doubleday Anchor Books, p. 363.

A similar argument is made by Andreski:

> The loyalty of the European peasant of not so long ago did not reach beyond the horizon of his native village. It was the enlargement of his contacts with the outside world, due mainly to industrialization and education, that enabled him to conceive the idea that he belonged to a nation comprising thousands of villages like his own. As "we" can be conceived only in contrast to "they", the peasant became nationally conscious when he came in contact with members of other nations, or at least apprehended their existence.[9b]

[9b] Stanislav Andreski, *Military Organization and Society*, London, 1968: Routledge and Kegan Paul, p. 89.

2. The mass army

Napoleon perfected the French state machinery, and knew how to use it for mobilizing and equipping huge armies. In the years between 1804 and 1813, a total of 2.400.000 men were called to arms. The human losses were enormous; already by 1806, some 1.600.000 men had died, part of them in battle, but the majority succumbing to various diseases and undernourishment. Mass conscription thus led to a hitherto fore unprecedented waste of men; crowded and unhygienic camps of course contributed heavily to the casualties. But even the military operations themselves involved losses which would have appeared irresponsible to the military commanders of the eighteenth century. Walter Millis has pointed out the carefulness and relative moderation that characterized military operations of that epoch;[10] in fact, the costs of recruiting, fitting out and maintaining armies made commanders very reluctant to

[10] Walter Millis, *Arms and Men*, New York, 1958: Mentor Books, pp. 16 ff.

accept a fight with the enemy. This is aptly demonstrated in the following statement by Moritz of Saxony:

> I am not in favor of battles and I am convinced that a clever general could wage war all of his life without ever having to get involved in battle. One has to engage in small encounters in order to make the enemy crumble away.[11]

This restrained view of warfare was quite alien to Napoleon. In contrast to the generals of the early nineteenth century, he could take advantage of the legal sanctions of the law of 1793, which permitted him to tap the large manpower resources of France. In addition he was able to draw on other sources as well. The Grand Army of 1812 counted around 500.000 men;[11b] only about half of these were Frenchmen, as substantiated by the figures for the number of battalions (279 French, 283 foreign) and cavalry squadrons (220 and 263, respectively).[12] Only a few thousands, primarily officers, non-commissioned officers and long-serving men, returned from the Russian campaign. Eleven-twelfths of those who did not come back had died from lack of sufficient clothing and feeding: one twelfth were battle casualties. Thus only a minority of the losses were in accordance with Napoleon's dictum that "troops are made to get killed".[13]

3. The division principle

As a result of the greatly increased size of the armies, military strategy and tactics underwent profound changes. No longer was it possible for one commander to direct and lead a whole army; and, more important, no longer was it necessary to keep the whole army together in order to insure maximum protection during operations. Also, a half-million army allowed more complicated operations and movements designed to confuse rather than confront the enemy.

The *division principle* took account of these facts. This invention in military organization is usually ascribed to Moritz of Saxony, who is reported to have used it on a few occasions during campaigns in the early eighteenth century. It was incorporated into the French army in 1793, and first put to large-scale use by Napoleon Bonaparte. Later the divisions were taken up by almost all European armies.

Basically, the army division was organized into 1–4 infantry brigades, 1–2 cavalry brigades, artillery, engineering, and support units.[14] The division was able to operate independently, due to its self-supporting organization and high firepower. (The Gribeauval artillery system gave the French divisions 4–, 8–, and 12– pound field artillery pieces with a firing range of 600 to 1.000 meters and a speed of two shots per minuts).

This delegation of command resembled the division of labor in industry, as Alfred Vagts has pointed out:

> /T/he division in the army compares with the *atelier concentré*, the gathering of many processes of manufacture under one roof with division of labor in an industrial plant. The military transformation accompanied the industrial revolution or evolution. From the standpoint of national economy, the *atelier concentré* in eighteenth-century France indicated the decentralization of control, or delegation of powers, except in the most important state affairs, in order to achieve higher efficiency.[15]

---

[11] Quoted in André Beaufre, *Modern strategi för krig och fred*, Stockholm 1966: Prisma, p. 58. The same view was exhibited by William Lloyd in his *Military Memoirs* (1781), of which Vagts writes:
The activity of the military genius Lloyd wants to see rationalized by mathematics and topography; topography determines the direction of marches, camps, and positions. The general who knows these things can direct war enterprises with geometrical precision and lead a continual war without ever getting onto the necessity of giving battle". (Vagts, *op. cit.*, p. 81)

[11b] This figure from a French source (*Armeernas världshistoria*, III, p. 97) appears to be a conservative estimate. Hobsbawm (*op. cit.*, p. 93) gives the strength of *la Grande Armée* to have been 700.000, of which 300.000 non-French.

[12] *Armeernas världshistoria*, part III, 1968, p. 97

[13] Vagts, *op.cit.*, p.127

[14] *Armeernas världshistoria*, part III, p. 72

[15] Vagts, *op.cit.*, p. 83–84

With this new invention in military organization, possibilities existed to coordinate the movements of several large army units, marching independently on to a pre-determined position where they could strike simultaneously against the enemy forces.

Napoleon prevented his adversary from anticipating the point where he would pull his forces together, and thereby blinded and paralyzed the adversary. Napoleon could surround the adversary when the latter remained motionless (as at Ulm) or, better still, go around him and take up positions on his lines of communication in order to force him to fight with reversed fronts (as at Jena).[16]

[16] Beaufre, *op.cit.*, p. 59

As is usual with military innovations, the enemies soon learnt to use the same technique. Also, one innovation often gives the impulse to — or requires — another. The process of organizational differentiation of the armed forces has its natural counterpart in the process of bureaucratization: the centralization of command, the creation of specialized functions for planning and supervision, and the emergence of a managerial elite. Although very little of this could have been clearly visible around 1815, it was only to take four more decades before the world was painfully aware of the impressive effectiveness of these new principles, as they were demonstrated by the Prussians. (Cf. below, sec. 7).

4. Industry and weapons production

Up to the last decade of the eighteenth century industrial mass-production of weapons was an unknown phenomenon. The most common firearm of those times, the muzzle-loaded musket, was handicrafted by gunsmiths in small shops or at government armories.[16b] Available figures for weapons production clearly show the difficulties of equipping mass armies before industrial production gained full headway. The government of the newly independent United States founded the Springfield Armory in order to furnish the forces of the republic with the necessary arms. However, even after five years in production, by 1799 it had barely reached an annual production of 5.000 muskets (some machinery was already employed, however with very limited effect). Twelve years later, through the combined efforts of the armories at Springfield and Harper's Ferry, plus some private production, the annual output had increased to 32.000.[17]

[16b] Cf. Lee Kennett, *The French Armies in the Seven Years War*, Durham, N. C: Duke University Press, 1967, pp. 114–115.

[17] Millis, *op.cit.*, p. 51

The development was far more spectacular in England and France. In 1791, the three French factories at Maubeuge, Charleville, and Saint-Etienne delivered 42.000 firearms (which ,however, was considered "totally insufficient" for the revolutionary armies). Estimates by 1814 show that the French army was then equipped with more than 2 million guns, 200.000 muskets and 200.000 pistols, plus 700.000 firearms taken as booty. Although no figures for annual production are available, it is obvious that weapons production must have expanded heavily around the turn of the century. In England, 3.1 million firearms were produced between 1803 and 1816, which makes a yearly average of about 238.000.[18]

[18] *Arméernas världshistoria*, part III, p. 84.

A prime reason for this expansion was the solution of the standardization problem, i.e. the problem of manufacturing weapons with interchangeable parts. This was first achieved in England (around 1800) and a few years later by Prieur in France.[19] In the United States, Eli Whitney, a nail manufacturer in Massachusetts, developed theories about substituting machinery for craft production.[20] He was looking for a suitable product for applying the new methods of assembly-line production: the government's demands for weapons was the perfect answer.

The history of weapons development after the industrial breakthrough at the beginning of the nineteenth century is well known. Increased capacity in producing more and lighter arms; more effective weapons — the rifled gun around 1850, the breech-loaded rifle around 1875, automatic weapons (such as the Gatling gun, the mitrailleuse, the machine-gun) toward the end of the century; and increased capacity for producing huge amounts of ammunition for hand- and artillery weapons.[21] Today a continuous production of war materiel in many countries has become a goal in itself, a goal about which capitalists and labor alike often demonstrate considerable agreement.[22]

5. Implications of increased firepower — the delegation of tactical command

Tendencies towards moderation was a prevalent characteristic of eighteenth century warfare. The royal armies were expensive and replacements difficult to obtain. The need for conducting campaigns as economically as possible led commanders to use long-drawn maneuvers and position warfare rather than direct battle. But when on occasions the armies engaged in battle, the losses were usually very high; therefore, military leaders tried to avoid such confrontations, and tended to prefer compromise peace treaties.

The weapons used by these armies contributed to this relative moderation. The musket was heavy, inconvenient, slow, and inexact. Rain and snow usually made it inoperable as a shooting weapon. Its firing range was about two hundred yards, but even at a distance of fifty yards the shot stood a generous risk of missing its target.

Much of eighteenth-century tactics were related to these characteristic of the principal weapon. To make maximum use of the muskets, the soldiers had to be kept closely together and the guns fired in volleys; the final attack was carried out with bayonets.

The attacker could evade the worst effects of the defensive volley by dispersion or by taking cover, but this meant that his bayonet charge would start in a scattered condition to arrive piecemeal, and so be easily disposed of by the serried mass. Consequently, the attack as well as the defense usually adopted a close-packed formation, accepting the horrible slaughter which this invited in order to conserve its own weight of fire and impact with the bayonet.[23]

Aside from the use of muskets in the initial stage of the battle, eighteenth-century tactics were not too dissimilar from that of ancient

[19] *Ibid.*, p. 84

[20] Millis, *op.cit.*, p. 52

[21] The armored warship and the aeroplane contributed to institutionalizing war production as a factor constantly increasing in importance in the economy of the industrialized countries. The airplane was of rather little importance to the combatants of the First World War, compared to its role in World War II. However, from this one should not infer that production was insignificant. On the contrary: in France, more than 50.000 planes and over 92.000 engines were built during the period of 1915–1918. Similarly, Germany produced about 48.000 planes during the war. In 1915, the French aeroplane factories employed 12.600 workers; at the armistice, the figure had risen to 186.000. At the end of the war, England had 34 plane factories, and Germany 33. (*Arméernas världshistoria*, part IV 1969, table 1).

[22] The *magnum opus* in the analysis of the functions of war and their substitutes is of course the allegorical *Report from Iron Mountain* (Swedish edition *Angående möjligheten och önskvärdheten av fred: Rapport från en forskargrupp vid Iron Mountain*, Stockholm 1968: Raben och Sjögren). Also compare John Kenneth Galbraith's *How to Control the Military* (Swedish edition, *Att hålla Pentagon i schack*, Stockholm 1969, Tema). The similarities between the anonymously published Iron Mountain report and Galbraith's pamphlet seem to be more than mere coincidences.

[23] Millis, *op.cit.*, p. 17

Greece or Rome. The main principle in both cases was concentration and massed attack; as in ancient warfare, cavalry was used for reconnaissance, flank attacks, and pursuit of escaping enemy troops. To some extent, the mass formations were made less dense by the introduction of linear tactics; this, however, was mainly a device for increasing the firepower and did not mean that the doctrines of mass attack were abandoned. Soldiers were not allowed to disperse or to take cover. Partly this was due to the principles of mass action themselves, but partly also from the conviction that lying down was unmanly behavior in war. It was also a safeguard against desertion, since the closely packed units made all disorderly behavior clearly visible to the officers.

But there were several examples to show that the principle of mass action by soldiers marching in close formation was becoming highly obsolete. The French revolutionary forces had begun the practice of taking cover behind trees or in ditches. This utilization of the terrain plus the dispersion of the soldiers made the old type of firing in volleys useless and, as Alfred Vagts says, the Allied army "found a ghost as their enemy and it could not be vanquished by their platoon fire".[24] The same experience was gone through by the British under Pakenham, advancing towards the forces led by Andrew Jackson at New Orleans on January 8, 1815. It is difficult to find a more effective demonstration of the confrontation of old and new tactics: the massive formation of the British was met by the fire from Jackson's men, protected behind cotton bales or in trenches. The comparative losses of that battle look almost unimaginable: 2 000 British troops killed, as against seven dead Americans.[25]

The massive formations continued for several decades yet to characterize infantry tactics, but increasingly it became obvious that the enhanced firepower of both the infantry and the artillery would force new principles of waging battle. The history of the American Civil War is to a large extent the history of armies that evermore went into the ground. As pictured by Walter Millis:

> The armies had begun to dig in almost from the beginning; as the war progressed they dug with increasing fervor. Against artillery and the long-range rifled muskets they used felled logs, hay bales, stone walls, railway cuts, road embankments and the shovel.... /Before Petersburg the armies/ arrived at a form of siege and trench warfare not outdone either by the trenches of northern France during the First World War or the bunkers of Korea. Naked flesh could no longer be asked or forced to stand up to the sheets of lead which the new weapons were pouring out.[26]

Increasingly it became necessary to disperse the troops and to utilize natural protection. This new tactic also meant increased delegation of authority. "The issue was no longer to move geometric blocks, batallions, regiments, and brigades in good order with simple command words, but to maneuver a long shooting line adjusting itself to the terrain, and which could not be reached by the generals' and colonels' voices".[27]

During the twentieth century this development was to continue, until at the time of the Korean war initiative was delegated down to small groups of soldiers. As Janowitz has pointed out, this has led to a shift in emphasis from authoritarian doctrines of leadership towards more

---

[24] Vagts, *op.cit.*, p. 113. A similar discussion is presented by Frederick Engels in *Anti-Düring*, Stockholm 1944: Arbetarkultur, pp. 227–239.

[25] Millis, *op.cit.*, pp. 62–63.

[26] *Ibid.*, p. 116

[27] *Armeernas världshistoria*, part III, pp. 172–177

manipulatory approaches.[28] This trend has been accompanied by another one: the increasing influence of the behavioral scientist armed with questionnaires, punch card equipment, and computer programs. Generals started to pay increased attention to factors like "morale" and "cohesiveness", now when such data were made readily available by sociologists and psychologists in military service.[29]

6. Technical specialization and military education

The tactical deployment of infantry and cavalry during the ancient wars in the Mediterranean and during the European wars up to the beginning of the sixteenth century required some elementary organizational insights; but war enterprises did not require technically trained personnel. The sixteenth century marks the emergence of the military engineer and fortification specialist, leading the work in erecting new, low-profile fortresses that could better stand up to artillery attacks than the old, high-walled castles. Two new military occupations came out of the sixteenth century into the seventeenth: the engineer, specialist in the construction and protection of defense sites; and the artillery officer, specialist in destroying them.

The first artillery pieces are known from the early fourteenth century.[30] The projectiles of these early guns were round stones and obviously did not do much harm. It was not until metal cannonballs came into use that the need arose to build other types of fortresses, with thicker walls and lower profiles.

It is important to stress the requirements that were made of the early military specialists. The engineers learnt how to use different building materials and to gauge the strength of various materials and structures which, later, was of great advantage in building roads and bridges. Artillery officers had to learn ballistics and mechanics, no doubt often by trial and error, but towards the end of the nineteenth century virtually from textbooks. French military writers — most prominently Bourcet, Guibert, and du Teil — gave important new contributions to military theory. At the military colleges in Valence and Auxonne, Bonaparte had studied Bourcet's treatise on offensive mountain warfare, Guibert's *Essai General de Tactique* (1772) and du Teil's writings on artillery tactics.[30b]

The breakthrough in military education took place during the last decade of the eighteenth and the first ten years of the nineteenth century. During this period a large number of military academies were founded. They were differentiated according to location and curricula, depending on whether they educated infantry/cavalry or artillery/engineering officers (the time of study for the latter categories usually being longer than that for the former) (see table 1). The technical orientation of the early schools is a notable characteristic, and technical schools were often established well before schools for the non-technical branches.[30c] The best examples are provided by the cases of Woolwich as against Sandhurst (1741 and 1802, respectively) and the Ecole Polytechnique as against Saint-Cyr (1794 and 1808).

[28] Morris Janowitz, *The Professional Soldier,* Glencoe, Illinois, 1960: Free Press, ch. 3.

[29] Cf. Samuel Stouffer et al., *The American Soldier,* Princeton, 1949–1950: Princeton University Press.

[30] *Arméernas världshistoria,* part II, pp. 102 ff.

[30b] George Rudé, *Det revolutionära Europa 1783–1815,* Stockholm 1967: Aldus, pp. 181–184

[30c] This point is brought out by Corelli Barnett, "The Education of Military Elites", *Journal of Contemporary History,* 1967, vol. II, no. 3, pp. 15–35.

TABLE 1
Military academies

Country	School/Location	Established in	Branch	Notes
Great Britain	Woolwich (Royal Mil. Academy)	1741	Artillery, engineering	
	Sandhurst	1802	Infantry, cavalry	
France	Ecole Polytechnique	1794	Artillery, engin.	
	Saint-Cyr	1808	Infantry, cavalry	2 years of study
Italy	Modena	1805	Artillery, engin.	
Prussia	Berlin Köningsberg Breslau	1810		3 years of study (art. and eng. officers continued studies for 1–2 years at Berlin or Munich)
Russia	S:t Petersburg	1832		2 1/2 years of study (emphasis on topography and geography)
USA	West Point	1802		(emphasis on technical education)
Chile	Escuela Militar	circa 1890		
Argentina	Escuela Superior de Guerra	1900		
Sweden	Karlberg	1792		

Sources: All countries except Chile: *Arméernas världshistoria,* part III, tables 9 and 15
Chile: Johnson, *The Military and Society in Latin America,* p. 70.
Argentina: Dario Canton, "Military Interventions in Argentina 1900–1966" in van Doorn, J.A.A (ed.), *Military Profession and Military Regimes,* The Hague, 1969: Mouton

In Latin America the Chilean Escuela Militar was to become the impulse in the start of military professionalization. It was established through German assistance, and was to serve as an emulative factor and educational institution also for the military establishments in other republics (notably Colombia, Venezuela, Paraguay, Nicaragua, and Ecuador).[31] Much of the sudden interest in military education in Latin American around the turn of the nineteenth century is explicable as a series of responses to the initiatives of Chile.[32]

Together with the build-up of the military educational institutions a shift occurred in the recruitment patterns. Especially at the artillery and engineering schools the cadets were recruited more and more from the burgher class. In Prussia the predominance of the Junkers in officer positions was massive; but the deviations from this pattern, however small, were represented by sons of the bourgeoisie concentrated in the "new" branches of the army. In 1806, the Prussian officer corps counted more than 7.000. Among them there were 695 nonnobles, usually belonging to the artillery and subsidiary branches of the service.[33] This influx of non-nobles to the technical branches alone seems to have been due to three factors: first, the energetically pursued policy of Frederick the Great to keep the bourgeois elements outside the officer corps (he "spent the last years of his life cleansing the officer corps of this objectionable material")[34]; second, the low prestige of the technical branches;[35] third, the fact that the technical branches required a certain intellectual level. Since no intellectual requirements hindered the access to infantry and cavalry positions for the Junkers, they preferred the traditional careers rather than the new ones; or, in other words, they had very little to win by seeking employment as artillery and engineering officers.

Another fact which emerges from the historical sources of the establishment of military education is the considerable exchange of services between the military and the economic and industrial institutions of civilian society. The technical education of officers to a large part seems to have contributed to the build-up of the infrastructure of the industrializing countries. For instance, regarding West Point, Millis holds that, around 1850,

> There was a strong interaction between civilian and military needs and between the engineers, inventors and factory managers who responded to both.
> It is strikingly indicated by the fact that for many years the United States Military Academy at West Point was virtually our only higher technical and scientific school; and that its graduates were so often found among the leaders of our scientific, engineering and industrial development. Jefferson had found the army officers useful as explorers and observers; the Navy was later on to use its officers in the same way — as in Charles Wilkes' explorations in the Pacific in the early /18/40's or Matthew Maury's pioneer work in oceanography — while the Army's Corps of Engineers was early to assume the responsibility for harbor developments and civilian internal improvements which it has since then retained. Many officers resigned the service to take posts in civilian industry.[36]

Robert P Parrott, an artillery man, became an executive in the West Point foundry, one of the country's leading iron-works. Another

---

[31] John J. Johnson, *The Military and Society in Latin America.* Stanford, California: Stanford University Press, 1964, pp. 70–71.

[32] *Ibid.*, p. 70

[33] Karl Demeter, *Das deutsche Heer und seine Offiziere*, Berlin, 1930, pp. 8–9

[34] Craig, *op.cit.*, p. 17

[35] Hans Speier, *Social Order and the Risks of War*, New York, 1952: George W. Stewart, pp. 234 ff

[36] Millis, *op.cit.*, pp. 82–83

graduate from West Point, John A. B. Dahlgren, helped in laying out the basic triangulation for the coastal charts of the United States. Josiah Gorgas, graduating from West Point in 1841, became Chief of Ordnance for the Confederacy and was largely responsible for the development of Confederate war industry.[37]

The graduates from the Military Academy played important roles in both the expansion of the frontier and in facilitating the consolidation of economic and industrial enterprises. Thus, the ties between the military and industrial establishments — so frequently commented upon by C. Wright Mills, John Kenneth Galbraith, Fred J. Cook, and other critics — is nothing specific to the twentieth century, although of course in absolute figures both the size of industrial production for military purposes and military-industrial personnel alliances have multiplied.[38]

In an similar way, Johnson emphasizes the part taken by the Latin American military in the development of the economies of the various republics.

Much the same case can be made for professionalization and modernization in pre-World War I Spanish America as was made after World War II for the militias as a modernizing force in the emerging countries of Africa and Asia. In both cases the armed forces were a part of a scientific vanguard in societies anxious to become technologically advanced but lacking a scientific intelligentsia.... The officers of the Argentine and Chilean armies, for instance, were in the forefronts of national groups who supported the development of modern communications and transportation. The railroad system of Uruguay — along with Argentina's the best in Latin America — was initiated and advanced by generals who became presidents.[39]

Thus, it seems justifiable to characterize the military profession as a "facilitating structure"[40] to the development of capitalism. The military profession could hardly have developed before the industrial revolution, since both the military and civilian sectors before that time needed technical specialists only to a limited degree. Further, neither was there a need for *organizational* specialists in the military until armies became very large and had to be subdivided into smaller units. The emergence of the military manager coincides with the developments in logistics, to which we now turn.

7. Logistics

The continued delegation of responsibility to lower-rank commanders could not take place without the simultaneous development of the possibilities of higher commanders in keeping contact with the dispersed units. Here the invention of the electric telegraph in 1844 came in very handy as did, later, the telephone and the radio. With these new means of communication, together with the rapid development of the railroads, the scene was open for a very fast acceleration in the field of logistics. The German General Staff was to make an appearance as the most ingenuous organization to master these innovations.[41]

Logistics is a branch of military science, beside strategy and tactics. According to one standard source, logistics includes "loading and shipping problems in all forms of transportation, the survey of transport facilities, the computation of space available as against tonnage and

[37] *Ibid.*, p. 83

[38] For a more recent criticism, see Adam Yarmolinský, "The Problem of Momentum" in Abram Chayes and Jerome B. Wiesner, eds., *ABM – An Evaluation of the Decision to Deploy an Antiballistic Missile System*, New York, 1969: Signet Books, p. 148.

[39] Johnson, *op.cit.*, p. 77.

[40] On "facilitating structures", see C. von Otter et al., *Om hjälpstrukturer och vårdideologi i kapitalistiska samhällen,* unp. paper, Dept. of Sociology, University of Uppsala (mimeo.), 1970.

[41] The American Civil War to a considerable extent involved fighting over logistical facilities, and the acts of war were often directed against waterways and railway lines. Thus, the North won an important strategic advantage by securing command of the Ohio-Mississippi system, cutting off the South's access to the material and manpower resources of Louisiana and Texas. And "with the fall of Atlanta in the autumn of 1864, the South lost the heart of her internal rail net". (Millis, *op.cit.*, p. 112).

manpower requirements, calculation of the time in which a given movement may take place and the time lag factor in all operations, and a knowledge of material assembly problems in relation to battle efficiency".[42] Thus, logistics is intimately tied to the development and status of transport techniques.

However, since the most important requirement of armies is the supply and replenishment of manpower, a large variety of services for personnel administrative functions have also developed beside the technical ones. To illustrate the extreme importance of logistic planning in modern armies, we may compare the logistical needs of the American revolutionary forces at the end of the eighteenth century to the highly diversified logistical functions in the nuclear-weapon equipped US Army of the mid-1950's. The rudimentary forces fighting against the English required for their existence mainly food, clothing, powder, lead, and guns (where the procurement of guns was the largest problem).[43] Logistical planning and organization was typically on an *ad hoc* basis and, in comparison with contemporary armies, highly primitive. At the time of the Korean war, the American army had its logistical needs met by a large variety of organizational set-ups divided into (a), administrative and (b), technical services. (Table 2). Although several of these branches have been long in existence, functions such as the operation of pipeline systems, electronic warfare, supply and maintenance of CBR weapons, and supply and maintenance of missiles reflect the vast expansion of occupational specialties within the military field that has taken place only relatively recently.

Successful employment of military forces hinges on the skillful utilization of logistics. The effectiveness of logistical planning was dramatically demonstrated to, and realized by, the peoples against whom Prussia turned her armies in the wars of 1866 (Austria) and 1870–71 (France). In 1857 Helmuth von Moltke, "der grosse Schweiger", took over the command of the German General Staff. He was able to draw upon the studies of the employment of railroads in warfare thad had been carried out by his immediate predecessor, von Reyher. Moltke's motto, "getrennt marschieren, vereint schlagen" which was to symbolize his fame as a strategic genius both to his contemporaries and successors, seems a very logical consequence of the developments in the military field up to his time: one, the differentiation of armies into divisions; two, the possibilities (enhanced by steam-powered transport) of moving various forces independently of each other to a pre-determined point of attack; and three, the centralized command, facilitated by the development of telecommunications.

According to Moltke's opinion, the strategically problematic position of Prussia could be compensated best by ensuring a high degree of mobility of the armed forces rather than through the strengthening of fortifications along the borders. Consequently, a special railroad command was established, and the troops were equipped with field telegraphs, portable tents, and transportable field kitchens.[44]

[42] John W. Barnes, "Logistics", *Collier's Encyclopedia*. New York, 1962.

[43] Millis, *op.cit.*, p. 33.

[44] Walter Görlitz, *Der deutsche Generalstab*, Frankfurt am Main, 1950: Verlag der Frankfurter Hefte, pp. 100–101.

TABLE 2

Logistical services in the US army

Administrative	Technical	Functions (examples)
Adjutant general	Chemical Corps	Supply of chemical, biological, and radiological (CBR) weapons. Maintenance of CBR supplies. Protection against, and decontamination of CBR effects on vital areas. Training of troops in CBR warfare.
Finance	Corps of Engineers	Construction and maintenance of all buildings, fixed refrigeration plants, utilities, and pipeline systems. Demolition. Insect and rodent control. Surveying and geodesy. Fire fighting and prevention. Camouflage. Administration of real estate.
Chaplain		
Inspector general	Medical Service	Supply of medical, dental, and veterinary equipment. Care of sick and wounded. Nutrition and sanitation. Food inspection. Physical examinations. Analysis of enemy biological agents.
Judge advocate	Ordnance Corps	Supply of missiles and conventional weapons, ammunitions, vehicles. Technical advice on safety procedures. Operation of ordnance installations.
Provost marshal	Quartermaster Corps	Supply of all quartermaster materiel, including equipment used in airborne and aerial delivery operations (parachutes, containers, etc). Operation of petroleum pipelines and product laboratories. Laundry, dry-cleaning, and disinfestation services. Operation of bath units, bakeries, commissaries, clothing sales stores, baggage warehouses, and effects depots.
Special services		
Troop and public information	Signal Corps	Service of communication, telemetering, television, and electronic warfare systems. Pigeon training. Cryptography. Photography.
	Transportation Corps	Transportation of Army property and personnel. Operation of inland waterways, military railways, and highway transport.

Source: John W. Barnes, "Logistics", *Collier's Encyclopedia*, 1962

The employment of steam-powered logistics, however, was only the most visible characteristic of the efforts of the General Staff. One of its main achievements was the development of a new concept of rational planning. Specially detached groups of officers worked on war plans applicable to most parts of the European and North Atlantic area. The First Division (1. Abteilung) worked on Sweden, Russia, Turkey, and Austria; the Second Division on Germany, Denmark, Italy, and Switzerland; and the Third Division on France, England, the Netherlands, Belgium, Spain, and the USA.[45]

[45] *Ibid.*, p. 100

Thus, with Moltke the doctrine and working methods of the General Staff were developed into a well-calculated system. A permanent organ with rationally distributed tasks had instructions to collect and systematize all geographical, military, and statistical information, and to work out, on this basis, the corresponding mobilization and transport plans.[46] Each individual soldier received instructions for travel to a specific point of assembly when mobilization orders were given. At these points, troops were equipped and received further transport instructions. Every military unit was put on alert according to a schedule that, in detail, prescribed the mobilization procedure: arrival of the soldiers, distribution of uniforms, weapons, ammunition, horses, wagons, food rations, etc. Rail transportation was carefully planned in advance, with detailed timetables and instructions for the composition and loading of the trains.[47]

[46] *Armeernas världshistoria*, part IV, p. 292

[47] *Ibid.*, p. 297

Within a relatively short time, the waging of war came to involve a considerable amount of office-work; and at the time of the First World War, the rattling of the typewriter had become equally ominous as that of the machine-gun for the lives of the soldiers. As for the officers, their roles had often developed into a mixture of commanding and managerial tasks.

8. The military manager

Some years ago, in *The Professional Soldier*, Morris Janowitz presented five "working hypotheses" concerning the development of the modern military establishment. One of these hypotheses pointed to the narrowing skill differential between military and civilian elites.[48] Taking account of the fact that personnel with "purely" military occupational specialties had fallen from 93.2 per cent in the Civil War to 28.8 per cent in the post-Korea US Army, Janowitz said:

[48] Janowitz, *op.cit.*, p. 9

> The new tasks of the military require that the professional officer develop more and more of the skills and orientations common to civilian administrators and civilian leaders. The narrowing difference in skill between military and civilian society is an outgrowth of the increasing concentration of technical specialists in the military. The men who perform such technical tasks have direct civilian equivalents: engineers, machine maintenance specialists, health service experts, logistic and personnel technicians.[49]

[49] *Ibid.*, loc. cit.

This focus on the narrowing skill differential between the military and civilian society emphasizes the integrationist (cooperative, "organismic") perspective, and overlooks some important implications of the expansion of military administrative and technical tasks. Thus, it is not only the case that some such tasks have civilian counterparts but, more importantly, the military establishment *as a whole* shows a notable similarity with civilian society. Military training and practice today gives the military elite the experience and managerial expertise to run something which is, with few functions excepted, a replica of civilian society. The military elite is trained in running a society with *a legal system* (in which many laws are instituted and obedience to them ensured through the mediation of the military profession itself); *an educational system; a communications system* (in many respects more complicated and effective than that of civilian society); *a transportation system* (operating railways, highways, pipelines, waterways, and air lines); *a health service system; an engineering system* (organizational and material reseources for constructing and maintaining buildings, roads, ports, etc); and a number of miscellaneous enterprises like laundries, bakeries, clothing sales stores, and the like (cf. table 2). The only functions which are not readily included into the military organizational structure are, first, production of base material like iron, coal, oil, textile fibres, and grain; and secondly, the reproduction of men. Were it not for these basic needs, military society could exist entirely on its own.

There is no other profession that gives its members systematic training in such a wide variety of important social functions. The potential of the military for "civic action" has also frequently been observed.[50] No wonder, then, that military men themselves often come to the same conclusions and decide to use their administrative and technical skills in tipping the civil-military balance to their favor when politicians have suggested and tried to carry out programs threatening military interests.

9. Total warfare

War has grown into an extremely complex undertaking, demanding detailed planning and long-term preparations. These preparations, although often construed by the military as defensive measures,[51] have became primary factors in the processes of mutual distrust between nations on which conflict spirals feed and through which prophecies become self-fulfilling.

The developments in communications made it possible for generals to lead the forces from a post far behind the lines (or, in nuclear warfare, from a subterranean control center); this has made war safer for the generals.[52] At the same time, other inventions have contributed in making it highly dangerous to civilians. The introduction of the aeroplane and air bombardment signified the birth of the concept of total warfare, equally directed against women, children, and the aged as well as against soldiers. Cities, villages, crops, industries, roads, railroads, and waterways all have become strategic targets.

Total warfare has made the military establishment more intimately intertwined with civilian defense institutions, and has brought centralized

[50] See Johnson, *op cit.*, p. 260 ff; for the parallel concept of "role expansion", see Moshe Lissak, "Modernization and Role-Expansion of the Military in Developing Countries", *Comparative Studies in Society and History*, vol. IX, No. 3, April, 1967. For more critical discussions of "civic action", see e.g. I.L. Horowitz, "The Military Elites", in S.M. Lipset and A. Solari, eds, *Elites in Latin America*, London 1967: Oxford University Press; and José Nun's contribution to Claudio Veliz, ed., *The Politics of Conformity*, London, 1967: Oxford University Press.

[51] Ironically, the foremost prophet of total warfare, general Erich Ludendorff, held the opinion that total war had no political cause, but was only a defensive act, partly brought about by the new means or warfare, and partly because of overpopulation. Cf. Speier, *op. cit.*, ch. 22 ("Ludendorff: The German Concept of Total War").

[52] After the fronts were frozen and the war of trenches had opened during World War I, a silent agreement seems to have existed between the combatants to leave each other's headquarters unmolested. The British Headquarters, for instance, remained for two and a half years in one place, "a spacious and comfortable castle", according to Vagts (*op. cit.*, p. 395)

organs for coordinating the combined security efforts of the state (for instance, the American National Security Council, and the Swedish total defense organization). Armament programs have created an increased interdependence between military and civilian economy. Total warfare has contributed to the blurring of the dividing line between strategic and political issues, as exhibited for instance by the debate on the ABM system in the United States during 1969-70. On the international level, military alliances such as NATO and the Warsaw Pact have meant a growing involvement of military officers in politics.

The demarcation line between military and political affairs has never been completely clear: sometimes, like in France, it has been precariously upheld; sometimes it has been an illusion for the masses, covering the true state of civil-military relations (e.g., Prussia); sometimes it has been consciously and deliberately violated (like in major parts of Africa, Asia, the Middle East, and Latin America). This has taken place concurrently with military professionalization. As professionalization has progressed, officers have increasingly become better trained in complex managerial tasks and have received extensive technical and organizational education. Against the background of the growing difficulty of separating the military and political sphere, it would seem a far-drawn conclusion that the military *because* of its professionalization might be expected to refrain from political intervention.

10. Summary and conclusions

(for a schematic representation, see Diagram II).

The emergence of the military profession in the form we know it today in the countries of Europe and North America is a consequence of several major social developments, such as

(a) *state centralization*, creating a rationale for national armies, recruited domestically rather than internationally, and facilitating (through national legislation and central bureaucracy) the mobilization of cheap military labor for mass armies;

(b) *industrialization* (and its corollaries, such as division of labor, solution of the standardization problem, and line production), laying the material foundations for the equipment of large armies, and establishing ties between the bourgeois class and the officer corps (replacing the traditional connections between officers and the royal and noble elites of the old regimes);

(c) the *decline of the nobility* as a recruitment source; and

(d) *technological innovations* (new means of transport and communications, weapons with high firepower, CBR weapons).

Consequences for professionalization$_1$ (p_1). Under the influence of these trends, the military establishment has

(1) grown considerably in *size;*

(2) become more and more technically and tactically *differentiated:*

(3) developed more *heterogeneous social recruitment* (partly as a consequence of (1) and (2)); and

(4) developed new *organizational forms* (permanent central staffs, logistical services, total defense institutions).

Consequences for professionalization$_2$ (p_2). Under the influence of trends such as (1) – (4), a number of new demands have been put on the officer corps.

(i) Complicated weapons, logistics, and new organizational forms have created a need for differentiated *education,* most clearly visible in the establishment of military academies around the turn of the nineteenth century.

(ii) The catastrophic prospects of nuclear warfare, the multinational defense alliances, and the fusion of military and civilian defense functions have brought about an increasing *political role-expansion* of military officers. Even minor military decisions may have repercussions for the escalation of international conflicts to the stage of nuclear exchange. The role-expansion has arisen as a natural counterpart to the growing overlap between strategic and political issues.

As a more indirect implication, the increasing heterogeneity of social recruitment has had the effect that corporateness among officers today is created on the basis of common professional values and goals. During earlier times, such corporateness was to a considerable degree automatically ensured through common social origins (nobility).

Also because of the heterogeneous social recruitment, military training today contains a considerable a degree of indoctrination of ethical rules (gentlemanly behavior, codes for proper behavior in uniform, various rules for paying respect to colleagues and superiors, etc). In contrast to their eighteenth-century colleagues, officer candidates today cannot be expected to have received the proper ethical socialization in their homes (as we shall see, in some countries considerable proportions come from workers' and farmers' homes): hence it becomes necessary to make these norms explicit during training, through officers' codes and manuals.

These elements of p_2 provide a justification for the use of the term "military mind" (see below, ch. 4). Professional indoctrination and controls, ensuring loyalty to professional goals and values, have emerged as substitutes for earlier forms of corporate cohesion, based on common social origins and similarities in family background. The "military mind" is at least partly a product of intra-professional socialization mechanisms, and a complement to the emergence of the military profession as a corps of specialized experts.

DIAGRAM II

Elements in military professionalization

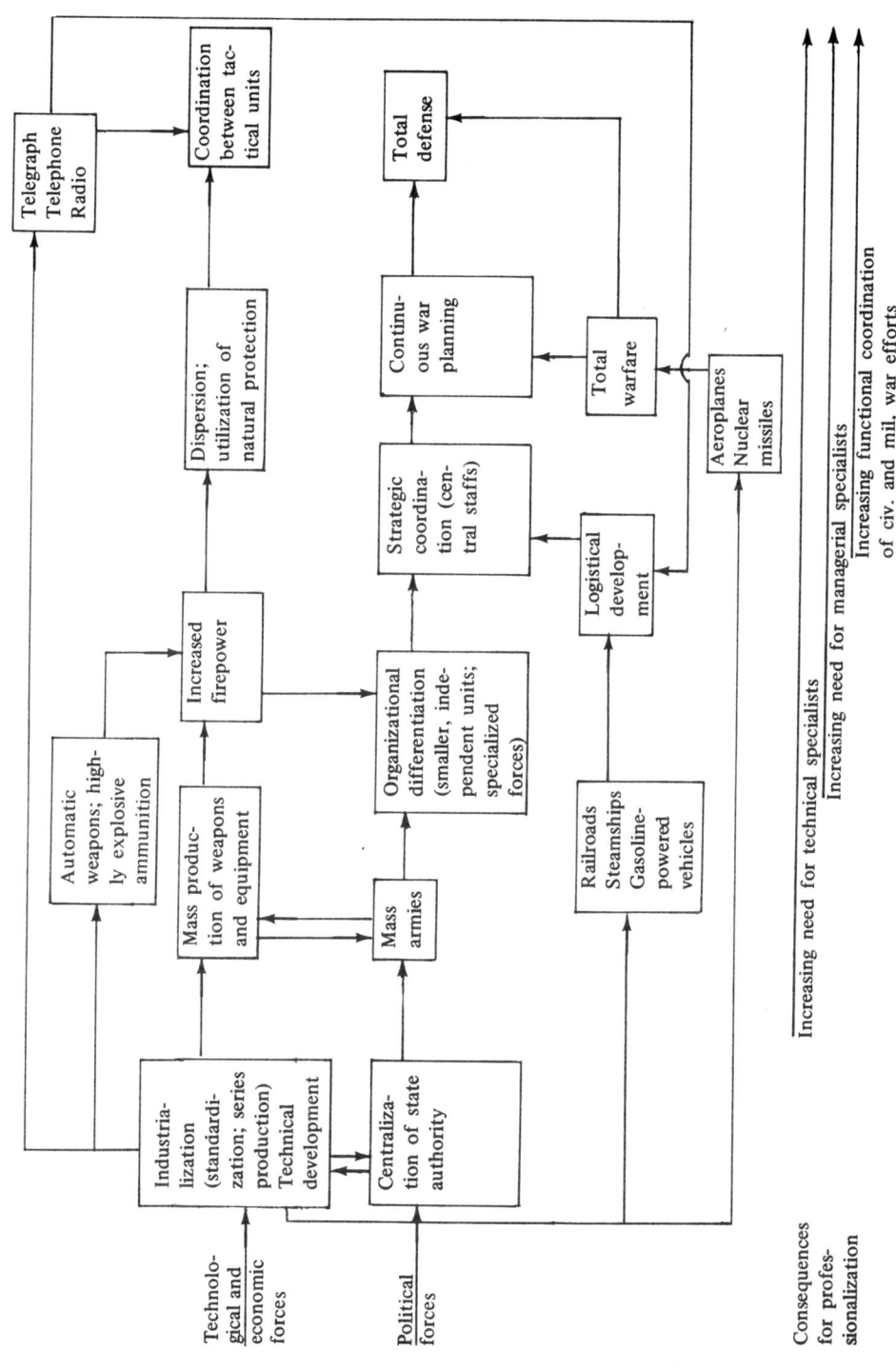

TWO

SOCIAL RECRUITMENT: INCREASING HETEROGENEITY

During the nineteenth and twentieth centuries, a considerable broadening of the social recruitment to the military profession has taken place. There are three major causes of this development: first, the decline of the nobility and the rise of the middle classes as the most important recruitment base; second, the vast enlargements of the armed forces of most countries, making any single stratum less likely to suffice as the sole provider of officers; and third, the new opportunities for education and social mobility offered by military educational institutions, attracting recruits from previously unprivileged strata.

The trend towards increased social representativeness reflects important socio-economic transformations as well as changes in the public image of the professional soldier. The managerial revolution has brought about a gradual shift of the officer role from the "heroic commander" to the technical and administrative expert. The latter role, requiring a number of skills transferable from the civilian sphere, presumably has been instrumental in toning down the traditional hesitation of the middle classes toward the military.

The increase in size of the armies since the turn of the eighteenth century, which in the long run has made the nobility and gentry an insufficient recruitment source, and the impact of military education, have contributed in diminishing the importance of ascriptive properties (such as noble origins and wealth) as criteria for military employment, so prevalent in the pre-revolutionary French and old Prussian armies. The growing importance of education and professional skill may give the impression that the military profession today takes its recruits fairly unbiasedly from all social strata. As we shall see, however, there are still important residues of the notion that certain classes are more fit for the role of officer than others. This may be exemplified by the existence of special provisions in favor of sons of military men in France and Spain, and by the deliberate policies of some Socialist countries to recruit mainly from the working and peasant classes.

In this chapter, I will first examine some theoretical problems involved in the application of ascriptive criteria. After that I will review empirical data on contemporary recruitment patterns in a number of countries; in most of them we will note a shift toward a more socially representative recruitment, although often a large proportion of the young officers still come from the lower upper or upper middle class.

1. Research on the social background of the military

Inquiry into the social origins of elite groups has typically been motivated by the assumption that data on social background may provide a background for the prediction of political behavior. Examples of such works are Lasswell's et al., *The Comparative Study of Elites* and C. Wright Mills's *The Power Elite*.[1]

In research on the military, an interest in social origins has been a prominent theme in such books as Janowitz's *The Professional Soldier*, Waldman's *The Goose-Step is Verboten* (dealing with the West German Bundeswehr), Demeter's *The German Officer Corps*, Vagts's *A History of Militarism*, and Craig's *The Politics of the Prussian Army*. These works vary with respect to ascriptive determinism, that is, the association postulated between social background (notably family origin) and political behavior. None of them, however, exhibits this tendency more clearly than Mosca's *Elementi di Scienza Politica*, which provides an almost ideal example of ascriptive determinism. Since Mosca's discussion may serve to demonstrate some of the limitations of this approach, a fairly detailed review seems merited.

Mosca points out how, in historical perspective, the non-intervening, a-political army has formed the exception rather than the rule in civil-military relations (we may in fact carry this proposition up to the present, in view of the frequency of armed take-overs in a great many developing countries).[2]

Mosca emphasizes the uniqueness of "the huge standing army that is a severe custodian of the law, is obedient to the orders of civil authority and has very little political influence...". His familiar approach to the question of why the European military during the nineteenth century avoided political interference[3] consists of relating the social origins of officers in the armies of the political democracies to the origins of the ruling elites (at the time of the publication of Mosca's *Elementi*[4] largely drawn from the aristocracy and the upper classes). The similarities in social background between the officer corps and the civilian elites becomes a primary factor in his explanation, exemplified by his remarks on the English military:

> The corps of English officers has always maintained a highly aristocratic character. The system of purchasing rankings held on in the English army down to 1870. In his *English Constitution,* Fischel justly notes that it is not the Mutiny Act that has kept the English army from becoming a tool for coups d'etat, but the fact that English officers belong by birth and sentiment to the classes that down to a few years ago were most largely represented in Parliament.

In a similar manner, the loyalty of the United States' army is explained by the class differences between high and low ranking military personnel, the outlooks and political preferences of the higher ranks presumably being more in accord with the outlooks of the ruling groups.

[1] Methodological problems in the study of social origins of elites are well treated in Geraint Parry, *Political Elites,* London, 1969: George Allen & Unwin, ch. 4.

[2] See e.g. S.E. Finer, *The Man on Horseback,* London, 1962: Pall Mall Press.

[3] The emphasis on non-interventionism, being strong in Mosca's work, is rather effectively disputed by Vagts who, while recognizing the fact that the European and American military have restrained themselves from military *coups,* nevertheless has ample comment on their interference in the political process. See Vagts, *op.cit.,* esp. chs. 9–11.

[4] Gaetano Mosca, *Elementi di Scienza Politica*. English translation, edited and revised by Arthur Livingston: *The Ruling Class,* New York and London, 1939: McGraw-Hill, ch. 9.

Mosca's thesis is based upon the assumption of a one-to-one correspondence between social origins and class indentifications, as illustrated by his proposition that "English officers belong *by birth and sentiment* to the classes most widely represented in Parliament".

We may question the general validity of Mosca's thesis for a variety of reasons. First, it should be noticed, he failed to take into account the case of Prussia, which constituted a very important deviation from his model. It is true that the Prussian military never engaged in a virtual coup d'etat; this does not mean, however, that the Prussian army was in any sense of the word "obedient to the orders of civil authority" nor indeed that it had "very little political influence". Gordon Craig's detailed historical description eloquently shows that there were other ways of intervention open to the Prussian military leaders than outright praetorianism.

To the political generals of the period after 1871 the thought of using a military *coup d'etat* to escape from their constitutional difficulties occured as frequently as it had to their predecessors in the 1860's. No serious attempt was made, however, to try this extreme and hazardous experiment. Instead, to defend themselves against what they considered to be the forces of revolution, the army chiefs relied on two principal lines of policy. In the first place, they progressively reorganized military administration in such a way as to withdraw the most vital military matters from the jurisdiction of the only person whom parliament could hold accountable for them — the War Minister — and to entrust them to constitutionally irresponsible agencies like the Military Cabinet and the General Staff. Secondly, they adopted a policy of officer selection which was deliberately designed to withhold commissions from persons with unorthodox social and political ideas and to maintain the officer corps as a bulwark of royal absolutism.[5]

Military intervention thus may take many other forms than the overthrow of governments; for example, the political dynamism of the German military was to continue in the well-known domination over the Kaiser by Hindenburg and Ludendorff.[6]

Second, during more recent times there has been a change of social recruitment to the military profession in the direction of greater middle class and even some working class representation. It appears, however, that *in spite of* this leveling the military have in large measure retained their traditionally elitist perspectives: political conservatism, emphasis on religous values, and stress on authority in interpersonal relations.

In some of these countries, notably Norway and Sweden, the officer corps has become one of the most "democratic" professions with regard to its social recruitment. But the attitudes and outlooks of the military have lagged markedly behind the ideological changes in the populations at large. Thus, although not more than 24 per cent of the Swedish officers in 1962 were of upper class origins, as many as 85 per cent indicated preferences for the Conservative party (supported by some 16 per cent of the total voting population in 1962). Hence, a broader social recruitment need not involve — as Mosca implied — stronger lower-class identifications.

Third, inferences from military social origins to military political behavior become less valid to the extent that officers are subject to

[5] Gordon Craig, *The Politics of the Prussian Army, 1640–1945*, Oxford, 1955: Clarendon Press, p. 218.

[6] See e.g. Vagts, *op. cit.*, p. 246.

pressures and expectations from other groups. The more a profession is exposed to such pressures, political or otherwise, the more it may be expected to choose strategies on the basis of their anticipated outcomes rather than on the basis of their consistency with social class norms and values.[7] Thus, the spectrum of options open to the group, and the values attached to each alternative, will determine actual behavior more than social background. If it is true that the military profession has developed a more political ethos (as Janowitz[8] maintains), then we would expect that social origin is less valid as a predictor of military intervention today than it might have been in Mosca's time.

Fourth, the process by which an occupational group gradually develops professional characteristics involves many features that tend to diminish the impact of social origin. Theoretical principles are codified into the curricula taught at the educational institutions of the profession. Professional journals are established; meeting and conferences are arranged. The professional society develops a tradition of its own. Sometimes, as in the military profession, there are devices like the rotation of personnel that serve to erode "localism" and the persistence of subgroup norms and values that could threaten those of the profession, as well as increasing the contact space for each professional member with other members. Thus, the profession commands a number of important means to create and reinforce solidarity to certain principles and interests. It certainly seems very natural that such principles and interest should come to prevail over what may remain of the professional member's family or class identifications.

To sum up, Mosca's model of a strong relationship between social background and political behavior appears dubious both because of some important deviant cases (such as Prussia), and because of the impact of intra-professional mechanisms that tend to counteract the influence of social origins. Thus, ascriptive determinism as an *assumption* about the true relations between social background and military political behavior seems inadequate. Hence, one might be tempted to leave aside the question of social recruitment as being theoretically unfruitful. However, ascription as underlying actual recruitment policies in several countries is still important; the fact that military and political leaders still *use* social origin as one discriminatory factor in the composition of the officer corps warrants continued interest in social recruitment. Let us turn, therefore, to a review of the research on social origins of officers.

2. Social recruitment — some data and interpretations

Janowitz has clearly demonstrated the decline in officer recruitment from the nobility and upper classes which took place in Northwestern Europe, Italy, and the United States during the nineteenth and the first half of the twentieth century.[9] He attributes this trend primarily to the increased need for manpower and the technologization of the military establishment, requiring a greater emphasis on intellectual and scholastic criteria for selection.[10]

[7] The general point in this paragraph is well born out by Reinhard Bendix and Seymor Martin Lipset in their article "Political Sociology", *Current Sociology*, vol. VI, 2, 79–98.

[8] *The Professional Soldier*, Glencoe, Illinois, 1960: Free Press, p. 12

[9] Janowitz, *op.cit.*, ch. 5

[10] *Ibid.*, p. 10. Note however, that the introduction of the artillery and the engineer corps into the continental armies was accompanied by a very low prestige for these "unestablished" branches. The breakthrough of bourgeois recruitment, which took place in them, was therefore probably due as much to the lack of prestige as to the need for "intellectual and scholastic" achievement criteria.

These findings have stimulated a number of replication studies outside the United States which have testified to similar trends toward social representativeness, for example, in Norway[11] and the Netherlands.[12] And for Latin America, Johnson notes a corresponding development, attributing it to the swelling of civilian bureaucracy and to educational opportunities:

> There would appear to be two principal reasons why more officers now come from the lower sectors than at any time since the midnineteenth century, when the armed forces were in near-total disgrace. First, economic and technological developments and the tremendous expansion of civilian bureaucracies have broadened the opportunities for the best-educated in the non-military areas. Second, public schools have given the lower classes the opportunity to receive the academic training needed to qualify for the military schools. Reasonably accurate figures show that in Ecuador 90 per cent of the cadets enter the military academies from the public schools....[13]

The recent data available for most countries show a rather pronounced middle-class dominance of the officer corps. However, the results defy unitary generalization; in fact social recruitment to the military appears to fall into four separate patterns. To demonstrate this, the figures in table 3 are suggestive.

The table lists data on social recruitment among samples of army officers (I) and army officer candidates (II). For eight countries, both types of figures are available; for four countries, only figures of type I; for two, only figures of type II. Methodology, size of sample, and occupational classifications in the various studies differ from each other, sometimes considerably. Some samples are randomly selected (USA, West Germany, Sweden),[14] some equal to the total population of officers (Poland, Czechoslovakia; no totals are given in these sources). Four studies give data on military academy classes graduating in different years (Norway, the Netherlands, France, Spain). For the rest, sampling procedures are either unreported or inadequately described.

As far as possible, occupational categories have been grouped together in order to maximize comparability. Obviously, however, comparisons between figures I and II *within* each country can be made with greater confidence than comparisons *between* countries, because of the relatively greater standardization of occupational categories.

I particularly want to stress that the figures are reported here on their own merits, and should be interpreted with caution. Hopefully, table 3 will serve to demonstrate the need for systematic, methodologically standardized cross-national studies.

In order to further condense table 3, I have grouped the three rightmost categories together (thus "professional", "business .." and "military officer" yelding an "elite" category) in order to compare them with the figures for the lower classes ("worker" and "farmer" added together). Computing the differences between these two summary categories will give us a simple index of elite domination in each officer corps (table 4, column 3).

---

[11] Kjellberg, F., "Some Cultural Aspects of the Military Profession", *European Journal of Sociology*, VI, 1965, 2, 283–293.

[12] van Doorn, J.A.A., "The Officer Corps: A Fusion of Profession and Organization", *European Journal of Sociology*, VI, 1965, 2, 262–282

[13] Johnson, John J., *The Military and Society in Latin America*, Stanford, 1964: Stanford University Press, p. 106.

[14] The Swedish officer sample contains personnel from the Navy, Coastal Artillery, and Air Force as well. Since the Navy is somewhat more exclusively recruited than the other categories, comparisons between categories I and II slightly exaggerate the degree of occupational mobility in the Army.

TABLE 3
Social origins of army officers (I) and army officer candidates (II)
Per cent

Country		Year	Father's occupation							Total	n
			Worker	Farmer	Crafts-man	White collar worker	Profes-sional	Business, managerial, landholder, tradesman	Officer	Other; no inform.		
United States	I	1950	5	10	–	11	34	29	11	–	100	140
	II	1960	19	–	–	13	25	15	25	3	100	765
Norway	I	1910–29	1	23	4	24	18	15	13	2	100	528
	II	1950–60	19	21	5	26	9	9	7	4	100	572
Sweden	I	1962	Lower: 13		Middle: 46			Upper: 24	17	–	100	462
	II	1956–58	Lower: 22		Middle: 52			Upper: 16	10	–	100	348
Nether-lands	I	1918–23	Lower: 1		Middle: 28			Upper: 41	26	4	100	222
	II	1948–51	Lower: 2		Middle: 38			Upper: 38	13	9	100	398
West Germany	I	1960	–	7	23	25	20	10	12	3	100	137
	II	1960	–	7	16	20	27	17	11	2	100	329
France	I	1937–39	3			9		14	25	5	100	600
	II	1954–58	5			6		12	30	14	100	350

Country	Period										
Poland	I 1958	51	34	3	—	— 10 —		—	2	100	no data
	II 1964	40	26	5	—	— 29 —		—	—	100	no data
Czecho-slovakia	I 1952	53	14	—	9	15	9	—	—	100	no data
	II 1966	66	7	—	3	15	2	3	4	100	no data
Great Britain	I 1959	—	—	—	—	11	30	44	15	100	36
Australia	I 1956	4	10	—	21	17	29	19	—	100	48
Ireland	II 1963	—	14	—	13	11	40	11	11	100	no data
Spain	II 1961–63	1	2	—	12	4	7	49	25	100	796
Argentina	I 1936–61	1	4	1	22	15	37	20	—	100	82
Chile	I 1952–64	1	19	—	8	24	22	26	—	100	37

Notes to table 1

United States. Source: Janowitz, *op.cit.*, p. 96, table 17.

Norway. Source: Kjellberg, *op.cit.* Army officer candidates during two periods. White collar includes 4 % schoolteachers (in both cohorts).

Sweden. Sources: Cohort I, Abrahamsson, B., *Anpassning och avgångsbenägenhet bland militärt befäl* (Adaptation to Work and Job-Quitting Proneness among Military Personnel), MPI rapport no 37, 1965 (Militärpsykologiska Institutet, Stockholm). Cohort II, Fråndén, O., "Notes on Mobility into and Out of the Swedish Officer Corps", in van Doorn, J.A.A., *Military Profession and Military Regimes: Commitments and Conflicts*, The Hague, 1969: Mouton. The lower class category consists essentially of workers. Figures for military officers estimated from Fråndén, *op.cit.*, tables 6 and 7.

Netherlands. Source: van Doorn, J.A.A., "The Officer Corps: A Fusion of Profession and Organization", *European Journal of Sociology*, VI, 1965, 2, 262–282. The figures are for army officer candidates during two periods. "Other" lists the percentage of father who are/were non-commissioned officers.

West Germany. Source: Waldman, E., *The Goose-Step is Verboten*, New York, 1964: Free Press, ch. 9. "Craftsman" includes skilled workers. "White collar" includes 16 % civil servants in cohort I, 12 % civil servants in cohort II.

France. Source: Girardet, R., and Thomas, J-P.H., "Problemes de Recruitment" in Girardet, R. (ed.), *La Crise Militaire Francaise, 1945–1962*, Paris, 1964: Librairie Armand Colin. Data (incomplete) for students at St Cyr (1937–39) and "division St Cyr" at the Ecole speciale militaire interarmes ESMIA (1954–58). "Worker" includes workers and low-level white collar personnel in industry and public functions. "White collar" = middle level white collar personnel in industry and public functions. "Business, managerial, etc" = managers and high level administrative personnel. Figures under "other" are for non-commissioned officers and gendarmes.

Poland. Source: Graczyk, J., "Social Promotion in the Polish People's Army", in van Doorn, J.A.A., ed., *Military Profession and Military Regimes*, The Hague, 1969: Mouton, tables 2 and 3.

Czechoslovakia. Source: Cvrcek, J., "Social Changes in the Officer Corps of the Czechoslovak People's Army", in van Doorn, *Military Profession and Military Regimes*, table 2 and p. 100. The 15 per cent in the "professional" column consist of "clerks and intelligentsia".

Great Britain. Source: Otley, C.B., "Militarism and the Social Affiliations of the British Army Elite", in van Doorn, *Armed Forces and Society*, The Hague, 1968: Mouton, pp. 84–108. Percentages computed from *ibid*, table 3.

Australia. Source: Encel, S., "The Study of Militarism in Australia", in van Doorn. *Armed Forces and Society*, pp. 127–147. Percentages computed from *ibid*, table 2. "White collar" includes 8 % schoolteachers.

Ireland. Source: Jackson, John A., "The Irish Army and the Constabulary Concept", in van Doorn, *Armed Forces and Society*, pp. 109–126, table 2. "Other" = members of the Gardai Siochona (police).

Spain. Source: Busquets Bragulat, J., *El Militar de Carrera en Espana*, Barcelona, 1967: Ediciones Ariel, ch. IV. "White collar" includes civil servants (6 %). "Other" includes non-commissioned officers (21 %).

Chile. Source: Hansen, R.A., *Military Culture and Organizational Decline: A Study of the Chilean Army*. Unpublished doctoral dissertation, Department of Sociology, University of California, Berkeley, 1967, table 40.

Argentina. Source: de Imaz, J.L., *Los Que Mandan*, Buenos Aires, 1964: Editorial Universitaria de Buenos Aires, p. 58. Sample of generals on active duty in the periods of 1936/41, 1946/51, and 1956/61. Percentages computed from available information (given for 33 per cent of total).

TABLE 4

Country	(1) Professional + Business... + Officer (per cent)	(2) Worker + Farmer (per cent)	(3) Elite domination (1) − (2)	(4) Rank
Argentina	72	5	67	2
Australia	65	14	51	5
Chile	72	20	52	4
Czechoslovakia	20	73	−53	14
France	⩾42	⩾ 5	37[1]	10
Great Britain	85	0	85	1
Ireland	62	14	48	7.5
Netherlands	51	2	49	6
Norway	25	40	−15	12
Poland	⩽29	66	−37[2]	13
Spain	60	3	57	3
Sweden	26	22	4	11
USA	65	19	46	9
W. Germany	55	7	48	7.5

Source: table 3

[1] Probably underestimated; data incomplete

[2] Probably overestimated, since figure in col. (1) also includes some white collar workers.

Given the last two columns in table 4, the countries fall into four patterns.

Pattern 1: elitist. Countries with ranks 1−3: Great Britain, Argentina, Spain (elite domination 85 to 57 per cent).

Pattern 2: upper class domination with some lower-class representation. Countries with ranks 4−10: Chile, Australia, the Netherlands, Ireland, West Germany, USA, France (elite domination 52 to 37 per cent).

Pattern 3: mixed, with about equal upper and lower class representation. Countries with ranks 11 and 12: Sweden, Norway (elite domination 4 and −15 per cent, respectively).

Pattern 4: working class dominance. Countries with ranks 13 and 14: Poland, Czechoslovakia (elite domination −37 and −53 per cent, respectively).

Commentary

Although the elite domination in *pattern 1* is very strong, it derives from somewhat different sources in the three countries. In Great Britain, the prevalence of the elite is due both to a high proportion of officers coming from the traditional base of the English establishment (professionals, landed gentry) and a very high proportion of military occupational inheritance (44 per cent).[15] At the other extreme is the case of Spain, with the highest percentage of officers' sons to be found among the countries in the table (49 per cent; note also that an

[15] The index of occupational inheritance is equal to the per cent of the men in an occupational category whose fathers were in the same category (inflow). The index of self-recruitment is the per cent of fathers whose sons continue in their occupational category (outflow). See e.g. Blau, P.M., and Duncan, O.D., *The American Occupational Structure*, New York, 1967: Wiley, p. 40.

additional 21 per cent included in the "other" category are sons of non-commissioned officers), but a low representation of other elite groups. Argentina falls somewhere in between these two cases, with only a moderate percentage of occupational inheritance, but a high proportion of officers coming from the professional, business and landholding classes.

In Spain, the high proportion of officer candidates coming from military families is to a large part explained by the existence of special provisions for this category, provisions that have a long and well-established tradition. At the establishment of the Academia General Militar in 1882, a number of articles stipulated different ages of admission for sons of officers and of civilians; stated that sons of officers were always to be given at least fifty per cent of the positions open; and gave special favors to "war orphans", i.e., sons of officers killed in battle. A similar provision assigned a special pension to sons of officers serving in the General Staff; these latter privileges were not accessible to sons of non-commissioned officers or farmers.

Although the more conspicuous of these self-containment stipulations today appear to have been removed, special benefits for "military orphans" still exist, together with different requirements for age of entry into the AGM. For instance, the maximum entry age for sons of military officers, NCO's and *guardias civiles* is 6–9 years higher than for sons of farmers and common soldiers, allowing the former categories to repeat their entry examinations several times over in case of failure.[16]

To continue along this line, let us look at *pattern 2* and start with the cases of France and Chile. In the French officer corps, as represented by the students at ESMIA (see footnote, table 3) in 1954–58, the elite domination is primarily due to the high proportion of sons of officers (30 per cent) and non-commissioned officers and gendarmes (14 per cent). In Chile, occupational inheritance is also considerable, although far less than in France (26 per cent). Other elite categories are proportionately stronger in Chile.

In both these countries, sons from military families have special advantages. The French military is to a large extent supplied with candidates from officer families through special preparatory military schools, open to children of French citizens who have fulfilled their military obligations. Among the students in the military academies in 1960 and 1961, about a quarter had previously attended the preparatory schools.[17] Thus, although the data on the French officer corps are incomplete, they suggest important differences between the social composition of the military in France and the majority of the other highly industrialized countries in table 3. Because of its heavy occupational inheritance, the French officer corps is more similar to the Spanish professional military.

Regarding Chile, Hansen points out how special advantages for sons of officers, for instance, guaranteed scholarships for the Military School, serve to encourage professional selfcontainment among the Chilean military.[18]

[16] Busquets Bragulat, J., *El Militar de Carrera en España*, Barcelona, 1967: Ediciones Ariel, pp. 100 and 155.

[17] R. Girardet & J.P.H. Thomas, "Problemes de Recrutement", in Girardet, R. (ed), *La Crise Militaire Francaise, 1945–1962*, Paris, 1964: Librairie Armand Colin, p. 47. The usefulness of the military career as a way of social ascent to children from large families is suggested by the fact that 28 per cent of the students at the division Saint-Cyr of ESMIA during the period of 1945–58 came from families with five or more children. (*Ibid.*, loc.cit.)

[18] Hansen, Roy A., *Military Culture and Organizational Decline: A Study of the Chilean Army*. Unp. doct. diss., Dept. of Sociology University of California, Berkeley, 1967, pp. 205–208

Among the countries in this group, the high proportion of officers with elite background is to some extent offset by some representation from the lower classes: for example, 19 per cent from the farmer categori in Chile and 14 per cent in Ireland; 19 per cent from the working class in the United States. With the exception of the USA, the samples are either small, and sampling procedures inadequately reported (Chile, Australia, Ireland); or show only very slight representation of the lower classes, as in the Netherlands and West Germany. The American trend towards greater working-class representation is not found in these two latter countries (compare cohorts I and II).

As for the Netherlands, there appears to have been a certain tendency toward greater social representativeness. The middle strata have increased their proportion, primarily as an effect of diminishing occupational inheritance. But it is certainly too much to say, as does van Doorn, that "these tendencies indicate the end of the officer profession as a privilege of a social elite".[19] Despite some changes in that direction, the Dutch officer corps is still relatively elitist in its social origins. Lower-class representation is almost totally absent.

[19] van Doorn, "The Officer Corps", p. 279

Neither are there any signs of the levelling of social recruitment to the Bundeswehr. The West Germany military is predominantly middle-class. The proportion of professionals, businessmen etc is markedly higher among the sample of officer cadets than that of Bundeswehr officers. The lower figures for the latter category may reflect the deliberate attempts during the 1950's at creating a corps of officers less elitist than that of the old Wehrmacht. For example, Waldman has described certain recruitment procedures at the establishment of the Bundeswehr which probably served this purpose. He says: "The problem of finding the right men and of placing them in the right positions was regarded as one of the main requisites for success in organizing German military units . . . . . A decisive factor in the reinstatement of a former officer or non-commissioned officer was the applicant's present outlook on life and his anticipated future attitude".[20] These procedures, while not explicitly using social origin as an index for selection, nevertheless probably led to a certain bias in favor of officers with less elitist origins. The figures in the table then may be assumed to reflect a swing away from this calculated equalitarianism and a tendency of consolidation of elite positions in the West German officer corps.

[20] Eric Waldman, *The Goose-step Is Verboten*, Glencoe, Ill., Free Press, 1964, p. 52.

Turning to *pattern 3*, we note a much lower proportion of the elite categories than in pattern 2. Norway and Sweden show some similarities to each other, although Norway's recruitment is markedly more lower-class based than Sweden's, due to a heavier representation of farmers' sons. It should be noted that the Norwegian officer corps has a long tradition of lower-class representation, judging from the figures for cohort I. Even as early as in the 1880's, as many as eleven per cent of officers graduating from the Norwegian Military Academy came from farmers' or workers' homes.[21] This figure might seem insignificant, but indicates all the same that the Norwegian officer corps must have been a

[21] Kjellberg, *op. cit.*, p. 284, table 1.

clearly exceptional case in a time when recruitment to most of the European armies was heavily upper class, and the middle strata had barely begun to get a foothold in the military.

In comparison, the Swedish officer corps is dominated by the middle classes. As in Norway, the lower class representation has increased over time (13 per cent of the "older" sample come from lower class homes, mostly workers' families, as compared to 22 per cent among Army cadets), but not conspicuously. In both countries occupational inheritance has diminished, reaching about ten per cent in the "younger" samples. This probably reflects the long period of peace in Scandinavia (for Norway and Denmark interrupted by the German occupation 1940–45), making the military profession less prestigeful and attractive as an occupational choice.

Today, the military in both Norway and Sweden is one of the most broadly socially representative professions, and military recruitment when compared with, for instance, that of law and medicine, is clearly much less elite-accentuated. Kjellberg reports on the Norwegian military: "Professional, managerial and business groups accounted in the last period (1950–60) for about twenty-five per cent of the recruitment of cadets to the Military Academy, while the corresponding figure among law and medical student at the University of Oslo in 1958–59 was still well above fifty per cent".[22] In Sweden in 1960–62, the faculties of law and medicine contained a proportion of, respectively, nine and eight per cent of the first-year students coming from workers' homes; the corresponding percentage among students at military colleges was 23 per cent. In 1956, 41 per cent of all students at civilian universities and colleges came from the upper class, compared to only 23 per cent at the Army college.[23] Differences have persisted since the middle of the 1940's, but have become more emphasized during the 1950's.

As already indicated, differences in social recruitment between the military and other established professions, as well as the decline of upper-class recruitment to the military and diminishing occupational inheritance, are probably in large part explained by the fact that a long peace period has contributed to the low prestige of the military. As we shall see later (ch. 9) few professions in Sweden today are attributed with so many negative stereotypes as the military. In addition, and probably as a consequence of these unfavorable images, Swedish military schools receive students who, from an intellectual standpoint, are clearly inferior to students entering other institutions of higher education; in fact, no educational institutions get students with lower average marks than the military colleges. Research from the 1930's and onwards shows this tendency to persist and even to have become more pronounced recently.[24] The long inactivity periods of the military establishment, and its negative intellectual image, have probably contributed to making a military career unattractive to Scandinavian students.[25]

In *pattern 4* we find the Socialist countries in Europe, represented by Poland and Czechoslovakia; the main feature of the social recruitment to

[22] *Loc.cit.*

[23] Fränden, O., "Notes on the Mobility into and out of the Swedish Officer Corps", in van Doorn, J.A.A. (ed.), *Military Profession and Military Regimes,* The Hague, 1969: Mouton, p. 110, table 2

[24] *Ibid.,* pp. 123–125

[25] No data are available for Denmark. It seems a reasonable assumption, however, that the results reported here on the whole would be replicated. Borup Nielsen gives some figures for reserve officers, showing 16 per cent recruited from farmers' homes and 13 per cent being sons of workers. Since reserve officers probably are somewhat more exclusively recruited than regular officers, these figures suggest a social composition of the Danish officer corps resembling that of Norway's. Steen Borup-Nielsen, *Reserveofficersundersögelsen 1964,* Copenhagen, 1966: Militaerpsykologisk Tjeneste (mimeo).

their armies appear, however, to be generalizable to some other countries as well (e.g., DDR and the Soviet Union).[26] The main finding, of course, is the massive "proletarization" of the officer corps: the representation of the working class (as defined by the original sources) is about 40 per cent in Poland, 66 per cent in Czechoslovakia, more than 80 per cent in DDR[27]; regarding the Soviet Union, it has been estimated that around 1930 a fourth of the officers were of working-class origin, and about fifty per cent from the farming population.[28]

This pattern is due mainly to (a) policies of special promotion of sons of workers and farmers, but probably also somewhat to (b) the low prestige of the military profession, especially among the intelligentsia.

Promotional policies. Wiatr states, in a comment on the post-war transformations of Polish society that "/t/he changes in political and economic relations, rapid economic growth, and to some extent heavy losses in the ranks of the Polish intelligentsia during the war made it imperative to promote various forms of social advancement of the lower classes".[29] Obviously, employment in the armed forces was one such form, contributing also to the political control of the army by the party structure. Thus, traditional mechanisms to restrict the access of the lower classes to military positions were employed to *promote* military enrolment for those very strata. "In pre-war Poland an officer was not supposed to marry a woman of lower status, and a President's Order (1937) stated that anyone disobeying this order might be dismissed from the services. In post-war Poland officers were officially encouraged to marry into the workers and poor peasants classes".[30] After 1956, this policy of proletarization was abandoned,[31] for reasons that have not been made clear.

The rapid mobility of workers' and farmers' sons seems to have been associated with problems of personal adjustment. As one Polish study shows, those who rapidly advanced to the officer corps from factories and farms exhibit a lower satisfaction with the profession than do those who have war-time military background or who belong to the youngest generation of officers, graduated from military academies and colleges. Wiatr attributes this to the relative loss of prestige of military positions, caused by the post-war lowered requirements for entry and promotion.

Prestige. The contemporary tightening up of the educational demands in the Czechoslovak army is apparently motivated by a policy of elevating the status of the military profession, once the immediate need for manpower had been met.[32] While in 1957 only 30 per cent of the officers had completed secondary education, the figure for 1966 was 72 per cent. During the same period, the capacity for internal education within the military establishment was greatly expanded. As may be expected, officers with a university education have much higher probability of promotion than others. It is likely that this relatively recent emphasis on scholastic achievement will in the future attract a greater proportion of officer aspirants from the intelligentsia, and may signify the end of "proletarization", as van Doorn suggests.[33]

[26] van Doorn, J.A.A., "Political Change and the Control of the Military", in *ibid., Military Profession and Military Regimes,* The Hague, 1969: Mouton, pp. 17–19

[27] *Ibid.,* p. 17

[28] *Ibid.,* p. 17

[29] Wiatr, J.J., "Military Professionalism and Transformations of Class Structure in Poland", in van Doorn, J.A.A., *Armed Forces and Society,* The Hague, 1968; Mouton, p. 234

[30] van Doorn, "Political Change . . .", p. 17

[31] Wiatr, "Military Professionalism . .", in van Doorn, ed., *Armed Forces and Society,* pp. 234–235

[32] Cvrcek, J., "Social Change in the Officer Corps of the Czechoslovak People's Army", in van Doorn, ed., *Military Profession . . .,* p. 101

[33] "Political Change . .", p. 24

In Poland, this already seems to have gained some way. The Polish People's Army is currently undergoing a relatively rapid transformation in the direction of greater inclusion of sons of the intelligentsia. The causes of this may be, (a) the general expansion of this category in Polish society, (b) a change in political strategy vis-a-vis the armed forces, (c) the abandonment of ascriptive rules like the intermarriage stipulations quoted above and, (d) efforts at intensifying officers' training programs and raising the proportion of officers with higher education.[34] Thus the attractiveness of a military job to men from homes with intellectual or white-collar traditions may be expected to rise. During the major part of the post-war period, recruitment from these strata also seems to have been retarded by comparatively low salaries: "the study of family budgets in the military community conducted in 1964 showed that the average income did not exceed that of lower officials or semiskilled and skilled workers, while it was considerably lower than the income of better educated and higher placed groups of the intelligentsia, and even lower than the income of very highly skilled workers."[35] Against the background of the changes indicated, it is possible that a shift is occurring in the prestige of the Polish military, away from the situation described by Sarapata a few years ago who pointed out that the most favorable opinions concerning the military profession "come from the rural population and from unskilled workers, the less favorable ones from the intelligentsia."[36]

[34] Graczyk, J., "Social Promotion in the Polish People's Army", in van Doorn, ed., *Military Profession..*, p. 91

[35] Wiatr, J.J., "Military Professionalism..", in van Doorn, *Armed Forces and Society*, p. 234

[36] Sarapata, A., "Pozycja wojskowego we w spdlczesnych spoleczeństwie polskim" ("The Military's Position in Polish Society"), quoted in Wiatr, *op.cit.*, p. 237

3. Prospects

To what extent may we expect the shifts in social recruitment to continue? Recruitment patterns of four of the countries in table 3 show signs of a gradual change toward the inclusion of more recruits from the middle and/or lower strata. These are Norway, Sweden, the United States, and the Netherlands. Two countries, i.e., Poland and Czechoslovakia, have had their recruitment base abruptly changed as a result of political transformations during the post-World War II period, and now seem to be swinging somewhat back to increasing middle-strata recruitment. Finally, for the rest of the countries, whenever data are available they do exhibit tendencies of upper-class consolidation (West Germany, France).

Is it reasonable to expect the tendencies of the first-mentioned four countries to continue, and perhaps to be parallelled in the recruitment patterns of industrialized countries in general? For a number of reasons, this seems rather doubtful.

First, the military career as a means of social ascent for the less privileged strata will diminish in importance to the extent that educational oppotunities and the prospects of occupational mobility *in general* become improved. For example, the Swedish educational reforms of the 1950's and 60's (state-financed support to university and high-school students, remodeling of the curricula in primary, secondary, and higher education, etc) have probably contributed in making the entrance into a non-military career as attractive, and often as easy, as into the military profession. In the Swedish Army War Academy class of 1925—27, the proportion of working-class cadets was 3 per cent; in 1956—58, 22 per cent; and in 1962, it had again dropped to 12 per cent.[37] In general, measures taken in order to improve the chances of social advancement of the lower strata will negatively affect the recruitment from these strata to the officer corps. Since all the industrialized states more or less adopt welfare state policies, this seems to be a factor impeding lower-class recruitment to the military.

Second, with continuing industrialization and urbanization, the farming population diminishes and will gradually cease to contribute its share of aspirants to the military. In the United States, the "younger" sample contains no cases coming from farmers' homes. In Norway, we may expect the decline of the rural recruitment base to be especially visible, in line with the developments which have taken place in Poland and Czechoslovakia (and no doubt also in Sweden, although detailed data are lacking). The rural base of the military profession has been strong in many countries. The working-class in Western European countries is largely antipathetic towards the military profession, in part because of the frequent occasions when the military has been used to combat striking workers,[37b] and in part because the military value-system, with its emphasis in conservatism, authority, and traditional virtues, stands in opposition to the values and ideologies of the workers' movements.

On the other hand, rural life coincides well with the military emphasis on physical prowess and outdoor life. As Alfred Vagts points out in his description of the nineteenth-century European armies, officers largely were sons of landowners; also, the soldiers to a very large extent came from rural areas. Sir John Fortescue, in his *History of the British Army*, pointed out that the army wanted "respectable, docile country lads, brought up by careful, thrifty parents in a decent cottage home".[38] More recent analyses also show the rural component in the armies' officer material still to be strong; we may note also the large proportion of sons of landholders in the contemporary British Army.[39] It is unlikely that the disappearance of the rural classes as a recruitment source in the Western industrialized countries will be fully compensated by increasing motivation among workers to enter the military profession.

Third, the propensity for occupational inheritance and self-recruitment among the professions is generally very strong, and, as shown by Blau and Duncan, in the United States higher than for any other occupational group.[40] Although their finding concern only the civilian labor-force it would appear that the tendency for the military profession is comparable. Furthermore, for at least two other reasons occupational inheritance will be a dominant feature in the officer corps of several countries: first, as we have seen, occupational inheritance is promoted by specially designed provisions for sons of officers, as in

[37] B. Abrahamsson, "The Ideology of an Elite", in van Doorn, J.A.A. (ed.), *Armed Forces and Society*, The Hague, 1968: Mouton, p. 81

[37b] For instance, at the beginning of industrialism in England troops were put to use in riots among the Gloucestershire weavers in 1756, among silk weavers in 1765, and against a sailors' strike in 1768. See Vagts, *A History of Militarism*, p. 105. For an account of more recent military-worker confrontations, see S. Klockare, *Svenska revolutionen* (The Swedish Revolution), Stockholm 1967: Prisma.
A Swedish official report in 1953 mentioned that "sentimental and traditional reasons", stemming from the attitudes of the working class against the military, has been a prime reason for the difficulty in recruiting working-class officer candidates, in spite of the fact that officer education is available to all categories of conscripts. *Befälsordningen vid infanteriet*, SOU 1953:28: Stockholm, 1953, p. 12. See also *Officersutbildningen inom armén m m*, SOU 1946:38, Stockholm 1946, ch. 1.

Spain, France, and Chile; second, since the officer corps in many countries has expanded during the period of 1935 to 1965, this further contributes to the capacity of the officer corps to supply its own recruits.[41] Occupational inheritance, then, may be assumed to be as strong as or stronger than the influx of the lower classes in the military establishment of several countries.

Fourth, we may add that the higher fertility rate of the lower classes will decrease as industrialization and urbanization increases.[42] This will lower their share of the total number of children destined to the labor force, and will probably also affect the proportion of lower-class aspirants to the military profession.

To put the discussion from the opposite perspective, we may expect *increasing* lower-class representations in the officer corps in countries where
(1) the military establishment provides opportunities for education and occupational mobility not easily found in the civilian sphere;
(2) the rural population is large and not rapidly diminishing;[43]
(3) deliberate policies of proletarization are employed;
(4) the fertility of the lower classes is high and not rapidly decreasing.
This better summarizes the situation in the new nations in Africa and Asia than in the industrialized countries. In all probability, during the next two or three decades, recruitment to the military professions in the industrialized countries will be characterized by the prevalence of the middle strata and, in some, by a marked self-recruitment. In the Socialist countries of Europe, recent policies of working class promotion may be expected to cease or at least to become de-emphasized. Educational improvements will contribute to bringing about professionalization and making the officer corps more autonomous, somewhat in line with the development in the Red Army during the 1920's.[44]

As we have seen, ascription is still an important factor in the recruitment policies of several countries (Spain, France, Chile, the Socialist countries), reflecting the influence of party ideology and/or military pressures to promote certain groups believed especially loyal to the state and/or the profession.

However, as previously pointed out, the very process of professionalization has the function of reducing or even eliminating the importance of ascriptive criteria. As recruitment to the military profession has ceased to rely on the nobility or gentry and the recruitment base has become broader and more diversified, military education, professional socialization, internal means of communucation, codes of ethic and other means of ensuring cohesiveness now bring about the uniformity of values that during earlier periods was provided by family and class influence. A preoccupation with the social recruitment of the military can therefore direct one's attention too much to features of the military establishment that are, at most, secondary to its autonomous, specialized, expertified, and politically influential role, deriving more from its properties as a profession than from its class characteristics. This has been especially well put by Francisco Kjellberg:

[38] Vagts, *op.cit.*, p. 156. Around 1900, German writers discussed the comparative value of urban and rural recruits, tending to reject the men from urban areas, allegedly because of bad health (in, for instance, L. Brentano and R. Kuczynski, *Die Grundlage der deutschen Wehrkraft* (1900); Kuczynski, *Ist die Landwirtschaft die wichtigste Grundlage der deutschen Wehrkraft?* (1905))

[39] C.B. Otley, "Militarism and the Social Affiliations of the British Army Elite", in van Doorn, *Armed Forces and Society*, pp. 84–108.

[40] Blau, P.M., and Duncan, O.D., *The American Occupational Structure*, New York, 1967: Wiley, table 2.2

[41] For example, the officer corps of the United States was 13 000 in 1935, 108 000 in 1965: the strength of the Soviet Army was 582 000 in 1935, 2 000 000 in 1965. In many other countries, the post-World War II army is approximately between 1 1/2 and 2 1/2 times the size of the pre-war force (source: *Statesman's Yearbook*, 1935 and 1965). It is likely that at least the USA and the Soviet Union during the coming ten to twenty years will self-recruit its officer corps to an extent which far exceeds the present ratio, due to the expansion quoted above and to the vast enlargements during the second World War (when the officer corps of the United States, for instance, counted some 830 000 men). Thus, the figures in table 1 indicating increasing self-recruitment in the

In terms of ideals and values the officer-corps seems to remain distinctly separated from the rest of the society. Its social origins might have had some significance in the past, but the contrast today between the uniformity within the group and the officers' social background strongly suggest that the group now derives its main characteristics from a particular culture, developing from occupational activities and quite irrespective of social composition.[45]

Summary and conclusions

1. By *ascriptive determinism* I mean the association postulated by some writers between social origins and political behavior of the military. This chapter opened with an attempt at evaluating Gaetano Mosca's thesis that armies in Europe and the United States have refrained from military intervention because of the correspondence between officers' social origins and those of civilian ruling elites. In general, predictions of political behavior of the military on the basis of social background were found lacking in validity for the following reasons:
(a) Mosca himself failed to take into account important deviant cases such as Prussia where, although military *coups* were absent, the military nevertheless intervened frequently in the political process;
(b) although recruitment to the military profession in many countries has become more socially representative — sometimes making the military the most "democratically" recruited of the professions — the officer corps seem in large measure to have retained their traditionally elitist outlooks;
(c) with the political role-expansion of officers, we may expect them to become more and more dependent in their actual behavior on a relatively wide spectrum of pressure groups, and adjusting their political actions to the contingencies of day-to-day bargaining and *ad hoc* alliances with other groups: in such situations, the impact of class origins will diminish;
(d) various features of the professionalization process can be expected to neutralize the importance of family background: conceivably, we may find relatively elitist perspectives even among middle- and lower class officers, to the extent that military professionalization fosters such perspectives.

2. Next, some comparative data on the social origins of army officers were reviewed. With due reservations about the quality of those data, the countries considered seem to fall into four patterns:
(a) elite domination (Great Britain, Argentina, and Spain);
(b) upper-class domination, some lower-class representation (Chile, Australia, the Netherlands, Ireland, West Germany, United States, and France);
(c) mixed (roughly equal upper and lower class representation: Norway and Sweden);
(d) working class dominance (Poland, Czechoslovakia).

3. The development in Norway, Sweden, the Netherlands, and the United States has been in the direction towards greater inclusion of middle- and lower-class recruits. For four reasons it is questionable that these tendencies will continue:

US Army will probably continue to rise. The same reasons may be advanced for a parallel development in the Soviet Union, and in other countries with rapidly expanding armed forces.

[42] Blau and Duncan, *op.cit.*, pp. 413–418.

[43] We expect the social composition of an occupation at least in some degree to reflect the *differentiation* of the society. (On the concept of societal differentiation, see Raymond Marsh, *Comparative Sociology*, New York, 1967: Harcourt, Brace, and World. Marsh has worked out an index of differentiation which is a composite measure of (a), the percentage of males in non-agricultural occupations and, (b), the gross energy consumption per capita. This index is computed for 114 contemporary national societies. See *ibid.*, appendix 1). Applied to the military profession, this means that there should be a smaller percentage of farmers' sons in the officer corps of highly industrialized countries than in the less industrialized ones.
Correlating Marsh's measure of societal differentiation and the proportion of farmers' sons for 12 of the countries in table 3 yields, as expected, a negative value, although not very strong (rho = −.37; for lack of data Sweden, the Netherlands, and France had to be excluded; had they been accessible, they would probably have given a still stronger negative correlation. The proportion of farmers' sons were taken from cohort I, except in the cases of Ireland and Spain).
The low correlation can partly be explained

(a) the military becomes less important as a means of social ascent when the possibilities of educational and occupational mobility in the society at large are improved;

(b) the farming population, which has been an important source for military recruitment, decreases;

(c) at the same time, the tendency towards occupational inheritance among the military can be expected to remain relatively high; and

(d) with continuing industrialization and urbanization, the fertility of the lower strata may be expected to decrease, lowering their share of the total number of children destined to the labor force.

4. As a general conclusion, the prediction of military political behavior seems to profit very little from studies of the social origins of officers. This is not without qualifications, however: to the extent that mechanisms of type 1(d) above are absent, we may expect the predictive power of social background indicators to be important. The less the degree of professionalization of a particular officer corps, the more we expect ascriptive criteria (such as class and tribal origins) to give clues to political behavior.

by the highly deviant case of Spain. Spanish society is relatively undifferentiated, with a large agricultural population (index 31.4 compared to, for instance, 45.8 for Poland and 58.0 for the Netherlands). In the Spanish officer corps, however, a very minute proportion come from farmers' homes.

[44] See Garthoff, R.L., "The military in Russia, 1861–1965", in van Doorn, J.A.A. (ed.) *Armed Forces and Society*, pp. 240–258; and P. Zhilin, "The Armed Forces of the Soviet State: Fifty Years of Experience in Military Construction" in van Doorn, *Military Profession and Military Regimes*, pp. 157–174

[45] Kjellberg, F., "Some Cultural Aspects of the Military Profession" p. 285

PART II

HOMOGENIZATION OF OUTLOOKS AND BEHAVIOR

(PROFESSIONALIZATION$_2$)

THREE

PROFESSIONAL SOCIALIZATION: THEORY, ETHICS, AND CORPORATENESS

1. Introduction[1]

It has long been recognized that activity within a particular expert field may have consequences for the values and outlooks held by the persons carrying out this activity. We expect the indoctrination of the particular goals and values of a profession to be effective to the extent that the profession is able to exercise hegemony over the content and intensity of the education of members, and to the extent that it can control their socialization without interference from other social groups. Because of its relative isolation from society, and the "total institution" character of military schools, the military here is at a certain advantage compared to most other professions.

A "professional mind" (to be discussed in further detail in the next chapter) is partly an effect of socialization processes; partly it also represents the effect of certain selective mechanisms. Leaving selection for later consideration, this chapter is devoted to three main elements in professional socialization: first, the learning of a body of theoretical doctrines; second, the implanting of ethical rules pertaining to the professional's behavior vis-a-vis his client, the public, and his collegues; and third, the feeling of corporateness and solidarity with other members.

Professional socialization will contribute to the crystallization of professional attitudes, i.e., a coherent system of outlooks facilitating the performance of professional tasks. The greater the degree of professional socialization, the more intensely felt and the more stable and consonant with each other we expect these attitudes to be. (We may also find crystallized sets of attitudes as a partial effect of selection processes; that is, the profession will tend to give higher priority to the selection and promotion of individuals with congenial rather than uncongenial outlooks. In the next chapter, this point will be discussed at somewhat greater length).

As already pointed out in ch. 1, the institutionalization of military education that took place around the turn of the eighteenth century meant an important break with the tradition of the officer as a part-time military leader, whose concepts of honor and corporateness stemmed

[1] The discussion in this chapter is adapted from the following sources: Julius Gould and William L. Kolb, *Dictionary of the Social Sciences*, New York, 1964, Free Press; two articles in S. Nosow and W.H. Form (eds), *Man, Work, and Society*, New York, 1962, Basic Books, namely, "The Emergence of Professions" by A.M. Carr-Saunders & P.A. Wilson, and "Attributes of a Profession", by E. Greenwood, Also H. M. Vollmer and D.L. Mills (eds.).*Professionalization*,Englewood Cliffs New Yersey, 1966, Prentice-Hall; and K.S. Lynn (ed), *The Professions in America*, Boston 1967, Beacon Press, especially the articles by Everett C. Hughes ("Professions") and Bernard Barber ("Some Problems in the Sociology of the Professions").
For the application of professionalization theory to the military, see e.g. S.P. Huntington, *The Soldier and the State*, Cambridge, Mass., 1957, Harvard University Press; *ibid.*, "The Military Profession", in Lynn, *op.cit.*; Morris Janowitz, *The Professional Soldier*, Glencoe, Illinois, 1960, Free Press; Charles H, Coates & Roland J.

primarily from his attachments to the nobility.

As the growing pressure from the bourgeois class began to be felt by the military establishments in Europe, the strong feudal traditions of the military gave way to values that were more compatible with bourgeois notions of achievement and education. The educational reforms within the military may be conceived as partial responses to these transformations.[2]

The new schools became primary instruments for the dissemination and consolidation of military theory, as developed primarily by French writers. But they also provided protected areas where new generations of the military profession could be brought up and corporateness could be reinforced in spite of tendencies of dissolution and "civilianization" through the influx of the bourgeois class. Their functions are largely the same today, although recruitment is still more heterogeneous.

2. Theory and professional autonomy

A professional theory consists of a set of doctrines that the members of the profession believe fundamental to their function and practice: in the military, the doctrines of strategy, tactics, and logistics and their appropriate subcategories. The theory is carried over to new generations through the mediation of schools, through printed and audiovisual media and through the day-to-day interaction between members. From time to time, the theory is supplied with new doctrines. Such revisions may be due to the failure of old methods and the apparent success of new ones (such as tirailleur tactics, trench warfare, or the armored columns of the Second World War); or they may be the effect of scholarly efforts to "look into the future", trying to determine what a future war will look like (cf. Kahn's *On Thermonuclear War*[3]).

Often a number of linkings with the "classics" are retained, for example works such as Clausewitz's *On War* and Jomini's *Traité des grandes opérations militaires*. Classical writings help the professional in defining this self-image as one who renders traditionally legitimated and "necessary" services to his client. Classical writers function as a sort of "model cases" in professional theory, providing the background for the professional's estimation of the originality of more recent theoretical contributions.

The theory gives the professional a certain authority vis-a-vis non-professionals, and provides the main basis for the autonomy of the expert group. As Everett C. Hughes has put it:

> Professionals *profess*. They profess to know better than others the nature of certain matters, and to know better than their clients what ails them or their affairs. This is the essence of the professional idea and the professional claim.[4]

The theoretical principles acquired during formal education and in the process of professional socialization form the **groundwork** for the application of professional expertise in the service of the client. The mastery of these principles is fundamental to professional autonomy. Conversely, the autonomy of the profession is threatened to the extent

Pellegrin, *Military Sociology*, University Park, Maryland, 1965, The Social Science Press; and two articles in the *European Journal of Sociology*, no. 2, 1965, namely, Jacques van Doorn, "The Officer Corps: A Fusion of Profession and Organization", and Francesco Kjellberg, "Some Cultural Aspects of the Military Profession".

[2] Cf. Alfred Vagts, *A History of Militarism*, London, 1959: Hollis and Carter, pp. 69 ff.

[3] For a sharp critique of Kahn's work, see James R. Newman, "Thermonuclear War", *Scientific American*, March 1961. Newman finds it hard to believe that Herman Kahn actually exists. "Doubts cross one's mind almost from the first page of this deplorable book: no one could write like this; no one could think like this. Perhaps the whole thing is a staff hoax in bad taste".

[4] "Professions", in Lynd, *op.cit.*, p. 2

that other groups make claims to know equally well or better how the problems in the particular issue area may be solved and advocate rival solutions.

From which groups, if any, comes the challenge to military hegemony? First, there are pacifist groups suggesting solutions to the problems of national security by means such as civil and non-violent resistance, non-cooperation, international agreements, and diplomacy.[5] An important force for the rational research and planning for general conflict control methods – i.e., not limited to military solutions – is represented by the various institutes of peace and conflict research, of which I shall have somewhat more to say later. Secondly, there are the "whiz kids" in the service of governments as a complement to the military group, i.e., civilian specialists in security policy.[6] To the extent that non-military defense analysts actually represent alternatives to military solutions, political strategies aiming at the control of the military may profit by improving the economic resources and influence of both these groups. As will be argued later, the institutionalization of policy alternatives requires the conscious strengthening of the policital power of pressure groups advocating such policy alternatives[7]

The emphasis on theory should not confuse the important fact that theory alone does not make a true professional. It is the combination of theoretical learning *and* the application of doctrines in practical work that makes the professional role rewarding (aside from its usually good economic prospects). Here, however, some problems arise for the military man. During peacetime the rewards inherent in the application of military theory are absent, and the professional soldier has to go through long periods of idleness. Attempts at partly solving this dilemma may involve the use of war games and simulated warfare, or voluntary enlistment in foreign armies or supranational military forces. Both these solutions are in balance with the profession's theoretical doctrines. However, the idle professional soldier may choose other ways. Professional training raises expectations that the principles which one has been taught will actually be used; hence, relative idleness in peacetime periods may cause officers to look for functional alternatives to "proper" military activity, activities in which his capacity of quick, determinate action may be utilized. This may partly explain the inverse relationship between external and internal conflict, for example in Latin America.[8] Conceivably, the high frequency of military coups may to a certain extent be due to the fact that the prospects of utilizing one's military expertise in actual warfare are slim; internal political action may therefore be an alternative fairly close at hand.

2.1 Schools, journals, conferences

The profession may be regarded as an institutional framwork for carrying certain theoretical principles in a particular issue area over to new generations. Informally, this takes place within the daily framework of interaction between the members. Formally, it is brought about by three important media: professional schools, journals, and symposia.

---

[5] See e.g. Adam Roberts (e.d.), *The Strategy of Civilian Defense.* Swedish edition *Civilmotståndets strategi,* Stockholm, 1969: Aldus.

[6] In Sweden the introduction of civilian experts in the planning of security policy has only recently taken place, partly against the resistance of upper military echelons. See Åke Ortmark, *De okända makthavarna.* Stockholm, 1970: Wahlström & Widstrand, pp. 177-78 and 224 ff. Some contributions to security policy and defense planning by non-military experts are given in Lennart Grape & Bengt-Christer Ysander, *Säkerhetspolitik och försvarsplanering,* Stockholm, 1967: SNS.

[7] See below, ch. 11 sec. 4.

[8] Stanislav Andreski, "Conservatism and Radicalism of the Military", *Archives Europeennes de Sociologie,* 1961, 1. Merle Kling, "Violence and Politics in Latin America", in P. Halmos, ed., *Latin American Sociological Studies,* The Sociological Review Monograph, University of Keele, 1967.

Schools. In general professional education is often integrated into civilian universities, as is the case with medicine and law in most countries. The military profession has been inclined on keeping its educational institutions isolated from civilian society and to circumscribe interaction between officer candidates and the civilian population. This speeds up the assimilation of military values and provides a functional basis for the process of "mortification of the self"[9] and the establishment of the new role. Furthermore, it may reduce the risk of young officer candidates being involved in social upheavals and political controversy. Military barracks are excellent means for protecting soldiers from the contagion of radical thought and from observing events threatening the status quo. When military education exists in conjunction with civilian institutions, as for example the American ROTC training system, this may call forth both civilian opposition and a potential decrease in political reliability of the soldiers.

Journals. The consolidation of the military profession in the early 1800's was accompanied by the foundation of a number of periodicals with military content, for instance the Prussian *Soldatenfreund*, the French *Spectateur militaire* and, later, *La France militaire,* the Prussian *Wehrzeitung* and *Militärwochenblatt,* and the *Army and Navy Journal.*[10] Contemporary military establishments usually issue journals of two kinds: one with contents of mainly theoretical interest and directed primarily to the higher levels of the military hierarchy (for instance, the *U.S Naval Institute Proceedings*), the other containing news and "professional gossip" about promotions, new uniform details, stories about battles in World War II, and the like, and are aimed primarily at the lower levels of the organization (for instance, the U. S. *Army Information Digest).*[11]

Symposia. A recurrent element in the life of a profession are meetings, symposia, and study visits, both nationally and internationally. As Alfred North Whitehead has pointed out, the professional community is to a large extent an international one,[12] and the communication between professionals of different nations is facilitated by their training in (usually) at least one foreign language. Thus, each professional institution "practices within its own nation, but its sources of life are world-wide. Thus loyalties stretch beyond sovereign states".[13]

While the basic ideology of the military profession is nationalistic, international contacts between officers in different countries are not totally lacking. Officers sometimes visit other nations, even outside military alliances of which their own country is a member. The need for international contacts is sometimes proclaimed also by military men in neutral nations, as in this quotation from an article by Swedish Admiral Stig H:son Ericson:

> A neutral professional of any kind cannot be without contacts with colleagues on the other side of the borders. Development of technology, defense materiel contracts, modern educational systems, the character of future war, pure courtesy visits, or other missions — all are reasons for foreign contacts being more important today than ever before.[14]

[9] E. Goffman, "The Characteristics of Total Institutions", in Etzioni, A., ed., *Complex Organizations — A Sociological Reader,* New York, 1964: Holt, Rinehart, & Winston, p. 318. Also compare G.D. Spindler, "The Military — A Systematic Analysis", *Social Forces,* vol. 27, 1958, pp. 83-88.

[10] Vagts, *A History of Militarism,* p. 304

[11] A further example is provided by the Swedish *Kungliga Krigsvetenskapsakademiens Handlingar och Tidskrift* (Transactions of the Royal Society of War) as against the more "popular" *Arménytt.*

[12] Law and social work do not fit this description too well, since their potential international character is hampered by their confinement to legal rules which are usually specific to each country (ch. Everett C. Hughes, "Professions", in Lynn, *op.cit.).* Thus, professionals trained in these specialties are particularly difficult to assimilate into a new culture when arriving as refugees or emigrés.

[13] Alfred North Whitehead, *Adventures of Ideas,* New York, 1933, Macmillan pp. 67-68.

[14] "Att vara svensk officer", *Fred och försvar under 60-talet,* Stockholm, 1964: Esselte p. 128

Thus, even in the military field with its security restrictions and emphasis on secrecy for national purposes, professional interests exert a pressure towards international communication.

3. Ethics

An important part of the professional culture is made up of ethical values, norms, and symbols aimed at regulating the behavior of the professional toward (a) his client, (b) the public, and (c) his colleagues. Such prescriptions fullfil a double function. First, as Greenwood points out, we may see these factors as protecting the surrounding society from abuses of the professional monopoly, as exemplified by the medical profession's Hippocratic Oath or the publishing codes of journalists' associations. The list may be made very long: as Greenwood emphasizes many occupations have such codes, and the professional ones are only more explicit, systematic, and binding. They possess "more altruistic overtones" and are more public service-oriented.[15]

Secondly, however, they also serve to define to others the professional sphere of interest, and to mold the public image of the professional. Thus press codes (such as those existing in Sweden) stipulating, say, that names of criminal convicts should not be publicized unless the court sentence is severe, over and above their internal effects also attempt to publicly define the role of the journalist as a responsible and trustworthy informer. It is as open question which function is the most dominant one; at any rate, frequent violations show that professional role *expectations* may differ markedly from actual behavior.

This has some applications also in the military field, particularly with regard to the military codes of fealty. While most officer corps have such codes, prescribing, for example, loyalty to the constitution, the history of civilmilitary relations shows a great many examples where such declarations of loyalty have been easily overcome. Constitutions are often vague enough to make them ideological servants of almost any group that is able to seize political power. Besides, loyalty to a constitution does not necessarily involve loyalty to a particular political leadership.[16]

It seems that writers of military oaths of allegiance have a certain bias against including paragraphs stipulating loyalty to the political leadership, and prefer formulations such as "the constitution", "the country", or national symbols such as the King or the flag. Thus, in the Swedish ceremony of "Krigsmans erinran" every soldier is solemnly reminded that the Swedish Defense acts for the preservation of "our peace and our independence" and the "security of the country". The text of the declaration does not mention any identifiable representative of "the country": supposedly, the soldier should be able to know what political institutions he defends.[17]

As a somewhat different example, consider the oath required of the armed forces of Honduras. According to the Charter's Article 321 it reads:

---

[15] Ernest Greenwood, "Attributes of a Profession", in Nosow and Form, *op.cit.*, p. 212-214

[16] Compare the statement by Mac Arthur, ch. 12, footnote 23.

[17] The older version of "Krigsmans erinran" abandoned in 1965, maintained that the soldier should "fear God and be faithful to the King"; at war he should "defend the King and the Fatherland with his life and blood". Besides this, no political concepts were used. *(Tjänstereglemente för krigsmakten,* Stockholm 1960: Försvarets kommandoexpediton, p. 2: 1-3. Supplement with new edition of "Krigsmans erinran", 1965).

> In my name and in the name of the Armed Forces of Honduras, I solemnly swear that we will not be instruments of oppression; that even if they come from superiors in rank we will not carry out orders that violate the letter or spirit of the Constitution; that we will defend the national sovereignty and integrity of our land; that we will respect the rights and liberties of the people, that we will maintain the apolitical and professional dignity of the Armed Forces, and that we will defend the suffrage of citizens and the alternation of the exercise of the presidency of the republic.[18]

Here reference is made to the executive branch of the government, although the primary emphasis is on traditional symbols of national sovereignty. It is probably difficult to find an example of an oath of allegiance which has less correspondence to the actual behavior of the armed forces. Honduras ranks among the absolute top among countries disturbed by military coups: during the 125 years between 1825 and 1950, the executive office changed hands 115 times.[19]

Military codes of honor sometimes serve indigenuous professional interests as much as, or even more than, protecting the public from abuses of professional activity. Thus the codes of Ecuador and Chile protect officers from arrest for civil crimes until after they have been tried by military courts and found guilty. In Chile, military courts even have jurisdiction over civilians who abuse the military as an institution or insult the national flag.[20]

Oaths and declarations of loyalty while being almost universally required of military men, seem to function more as a sort of label of apoliticality rather than actual impediments to military political action. They should be regarded primarily as protective and public relations devices for the profession; they seem to be insignificant in protecting civilian authorities from being overthrown by the military.

3.1 The concept of client

Professional practice is a dual role relationship: the professional cannot exist without a *client* of some sort. The concept of client calls for some discussion.

Traditionally, the clientele of the professions is often described as a plurality of individuals. This notion, typically based on the professions of medicine and law, has tended to become invalidated because of the growth in number and scope of organizations in modern society. The common picture of professionals is that of persons working in *freie Berufe*. But modern large-scale organizations have understood the gains in recruiting experts for a broad range of functions, and the personnel lists of such organizations now often include, for example

(a) lawyers to assist in drawing up contracts, and for counseling in labor legislation, insurance, and social security matters;
(b) engineers for the study and solution of various techno-mechanical problems;
(c) architects for the design of new products, industrial plants, workers' housing areas, etc; and
(d) physicians for company health services.[21]

---

[18] John J. Johnson, *Military and Society in Latin America*, Stanford, California: Stanford University Press, 1964 p. 161

[19] *Ibid.*, p. 5

[20] *Ibid.*, p. 113

[21] These trends have been strongly opposed by the American Medical Association. See D.R. Hyde, et al., "The American Medical Association: Power, Purpose, and Politics in Organized Medicine", *Yale Law Journal*, vol. 63, 1954, no. 7.

Thus, in the organizational setting the client of the professional is the organization itself rather than single individuals, although of course in daily business the organization is always represented by its officials. Similarly, the client of the military professional is society itself, as represented by civilian officials and politicians.

The fusion of profession and organization is not unique to the military although few other professional groups are tied equally strong to one particular organization with one and only one hierarchy of leadership. The Catholic church seems to present a fairly parallel case, although it is less nationally restricted than the military. Similarly, universities and other institutions of higher learning form organizational settings for teachers and scholars;[22] each university, however, is relatively independent of the others and represents one of several career alternatives to the individual scholar. If an officer fails in his career, however, he is bound to stay in his rank or to resign from service. Thus, although the process of "professionalization" of organizations tends to accelerate,[23] most experts in large organizations have a far less dependent position vis-a-vis their employers than the military men vis-a-vis the military organization; indeed, the latter is not likely to regard his organization primarily as an employer, but rather as an efficient instrument for coordinating the professional activity of himself and his colleagues.[24]

A common source of conflict in a joint organizational-professional setting is caused by the collision of organizational demands and the professionals' aspirations to independence. Organization professionals frequently have to face decisions whether to subordinate themselves under an official occupying an executive position — and thus surrender professional expertise to "layman" authority — or to maintain the claims to independence, at the possible cost of organizational disruption and perhaps the damaging of their careers. "Whereas professions find the pattern of 'colleague control' most suitable, the required pattern of authority for formal organizations is 'superordinate control'. The former consists of control by peers, the latter of control by superiors. As a result of these different types of required authority, it is inevitable that there be a certain amount of strain when professional roles confront organizational necessities."[25]

Since at most levels of the military organization "superiors" usually are also "professional peers", this type of strain can be assumed to be less prevalent in the military than in many other professional/organizational settings. However, an obvious exception to this is found at the highest level of the military establishment, where military commanders confront their professional views with those of government officials who usually have to balance military requests against requests from other pressure-groups as well. On these levels questions of the proper line of demarcation between military and political matters often become crucial. At lower levels, matters of professional concern are relatively easy to define as "purely military", while questions such as the deployment of new weapons programs, manpower expansion, military-industrial and

[22] Some of the most important men of science have been reluctant to subordinate themselves under the jurisdiction of the university. "Bacon, Harvey, Descartes, Galileo, Leibniz, wanted governmental patronage, or protection, more than university colleagues, mostly reactionary. When Luther, Descartes, Galileo, or Leibniz shifted his residence, it was not to find a better university, but a more suitable government — a Duke who would protect, a Prince who would pay, or a Dutch Republic which would not ask questions". A.N. Whitehead, *The Adventure of Ideas*, New York, 193: Macmillan. P.66

[23] Cf. Harold L. Wilensky, "The Professionalization of Everyone?", *American Journal of Sociology*, 1964, esp. p. 146; and Jacques van Doorn. "The Officer Corps: A Fusion of Profession and Organization", *European Journal of Sociology*, no. 2, 1965, pp. 262-265.

[24] See also ch. 11, sec, 3.1

[25] Bernard Barber, *op.cit.*, p. 25

military-scientific co-operation usually border on some aspect of domestic and foreign policy. The theory of civil-military relations in Western democracies contains an inherent contradiction, since the state is supposed to be *both client and superior* at the same time. The client is assumed to follow the advice of the professional expert; the superior is assumed to be the ultimate judge of what "treatment" is the most appropriate, and thus at times disapprove of the professional's recommendations. (These potential strains will be attenuated to the extent that the state is willing to make a broad definition of the military's jurisdiction, and/or to the extent that the military group itself occupies important positions in the councils of government).

As Huntington emphasizes, "the skill of the physician is diagnosis and treatment; his responsibility is the health of his clients. The skill of the officer is the management of violence; his responsibility is the military security of his client, society".[26] Professional advice, in peacetime, and military operations in war, are directed towards what the military leadership perceives as being in the true interests of the nation.

The emphasis is on *perceive*. Differences between the military, on the one hand, and governmental and other civilian groups, on the other, often occur with regard to what constitutes the "true" interests of the nation. Many of the problems concerning civilian control, and much controversy over the military budget and national security matters stem from the differences in risk perceptions between the military establishment and its client, society. Much more than in other professional-client relationships, a certain degree of distrust on part of the client is, and must be, institutionalized.[27] Hughes has stated that "since the professional does profess, he asks that he be trusted. The client is not a true judge of the value of the service he receives . . . ."[28] But unless the state chooses to abandon civilian control altogether, a strain always has to prevail between it and the military, precisely because the state as client must always be the ultimate judge of the quality of advice given by military experts. The military claims to the superiority of military expertise ("the proposition that in all service-related matters the professional group is infinitely wiser than the laity")[29] are — in theory, although not always in practice — counteracted by claims from the client/society that in all disputes the *client* is supposed to know better what is best for him. This is a basic principle in the democratic theory of civil supremacy; and it constitutes an almost perfect antithesis to the classical notion of the professional man as an independent practioner.

[26] Huntington, *The Soldier and the State* pp. 14-15

[27] That is, according to the theory of military subordination under civilian authority. For various reasons, however, the validity of this theory tends to become eroded; see esp. ch. 11.
[28] Hughes, *op.cit.*, pp. 2–3
[29] Greenwood, *op.cit.*, p. 215

3.2 Relations to the public

Although the most important professional—client relationships take place on the higher levels of the military organization, a number of rules also exist for regulating the behavior of soldiers in their dealings with civilians. Physicians and lawyers have codes of conduct prescribing, among other things, anonymity of the client, referral of the client to better qualified colleagues if necessary, and prohibiting advertising. A highly visible element in military codes of conduct is the emphasis on gentlemanly behavior, partly a derivation from chivalry traditions, partly reflecting the fact that the uniform sets the officer aside from the civilian community. For example, the Swedish Armed Services regulations point out:

> The soldier shall observe a conduct honoring his service and his uniform. He shall correctly address persons with whom he meets within and without the service.
> The soldier's speech shall be honest. Swear words and coarse language must not be used.[30]

And an instruction booklet for Swedish Army cadets lists the following instructions for social occasions:

> At a dance one should dance with as many ladies as possible. It is not courteous to the host and hostess nor to other ladies to conspicuously concentrate on dancing with only one or a few of the ladies or not to dance at all. Dance with style and avoid exaggerations. *All "physical intimacies" with the partner during the dance are prohibited. A cadet or an officer in uniform does not dance cheek-to-cheek even with his wife or fiance'e.*[31]

Such role prescriptions take on increased importance to the extent that cadets are recruited from social strata where norms against informal behavior on formal occasions have not been indoctrinated during early socialization.

Further examples of codes regulating the behaviour of officers vis-a-vis the public could be given,[32] and the reader may find citations of similar rules in, for instance, the American Army elsewhere.[33] As a matter of fact, the military appears to be more regulated by explicit rules of conduct than most other professions, perhaps with the exception of the clergy.

3.3 Colleagues

A second important group of specific ethical rules pertains to behavior towards colleagues. The military profession, again more than most others, strongly emphasizes cooperation, comradeship, and group cohesion. "In its contemporary form a major aspect of military honor comprises a sense of brotherhood and intense group loyalty," as Janowitz has expressed it.[34] And it is evident that, in terms of group loyalty and sense of brotherhood, civilian occupations compare unfavorably with the military profession in the opinions of officers. In a survey among the Swedish military in 1962, only 3 per cent of the officers, and 9 per cent of the lower-rank personnel considered potential civilian occupations to be better than the military with regard to colleague relationships. This may be compared to, for instance, opinions of promotion chances, which 50 per cent of the officers and 60 per cent of the lower-rank personnel regarded as more favorable in a civilian occupation.[35]

Janowitz has pointed out the function of certain symbols, such as the salute, in reinforcing the togetherness among the military. "One gets the

---

[30] *Tjänstereglemente för krigsmakten*, ch. 6, paragraphs 10 and 11.

[31] *Uppträdande utom tjänsten. Några råd och anvisningar för Kungl. Krigsskolans kadetter och arméns yngre officerare.* Stockholm, 1963: Kungl. Krigsskolan (mimeo), p. 26. Italics in orginal.

[32] The aforementioned booklet with rules for young officers also includes instructions for the first meeting with superordinate colleagues ("Sir, second lieutenant Lundborg asks your permission to introduce himself"); about how to shake hands properly ("a handshake should be firm, but neither too hard nor too loose"); about the proper use of titles; about correct and polite behavior towards superordinate officers ("politeness should be shown in a natural way and not – as is often the case – in noisy competition among junior officers"); about smoking ("a cadet should not smoke in the street"); about the proper design of business cards ("of moderate form and size, printed on evenly cut, white, unpatterned hard paper"); about when and how to visit the homes of senior officers ("if the person whom one wants to visit is not at home ... one puts his business card, folded in the upper right hand corner, in the mailbox"); about the paying of gambling debts, etc.

[33] Janowitz, *The Professional Soldier*, ch. 9

[34] Janowitz, *The Prof. Soldier*, pp. 220-21.

[35] Bengt Abrahamsson, *Anpassning och avgångsbenägenhet*

impression that the salute has been selected as a symbol of opposition to civilianizing trends, and is therefore maintained with determination. While it has become almost automatic, and in a sense peripheral to consciousness, it is still laden with powerful meaning".[36]

And an official Swedish military report on morale and discipline emphasizes that: "The military salute does not lack a value as a sign of cohesion, even if it is formally prescribed. For the majority of the conscripts the idea of solidarity probably has been regarded as having a positive effect".[37]

The internal loyalty and solidarity probably contributes to the effectiveness of the profession in carrying out its task, as Janowitz suggests.[38] However, the strong cohesion may also contribute to isolating the military from the broader context of the society; it may counteract factors promoting social change and may obstruct the re-definition of national priorities. Social cohesion is an important power resource in the corporate dealings with political authorities, since it functions as a check on political tactics of *divide et impera*.

Ethical norms and codes of conduct are not particularly effective by themselves alone, of course; they become so only when they are supported by and indoctrinated through the mediation of older officers. The military's relatively isolated situation also contributes to the reinforcement of group solidarity; and attempts by the civilian society to gain influence over military training and education — for instance by instituting schools which enable men from the ranks to qualify as officer candidates — are often resisted as illegitimate intrusion into professional matters.

4. Corporateness

The corporate spirit of the military profession is the result of a number of factors:

1. The long period of formal education during which friendship bonds are established, and the common theoretical and ethical basis for the profession is internalized;
2. Codes of conduct, rituals, and symbols, often strongly related to internal traditions, and perceived as being functional for the solidarity and cohesion of the officer corps;
3. The existence of media of communication with highly specialized contents, such as scientific journals, periodicals for inside information and "professional gossip", congresses, conferences, and study visits;
4. The system of rotation between staff and command functions (which may be seen as one way of adjusting to the military dilemma of reconciling the conflicting roles of military manager and heroic commander),[39] furthering widespread contacts among members of the profession;
5. Professional rewards such as promotions, medals, and honorary

bland militärt befäl, Stockholm, 1965: Militärpsykologiska institutet p. 138. The respondents were asked to compare their military job with a civilian occupation they might get with their present education.

[36] Janowitz, *op.cit.,* p. 221

[37] *Krigsmaktens anda och ordning.* Överbefälhavarens yttrande och förhållande mellan befäl och meniga. Stockholm, 1947.

[38] Janowitz, *op.cit.,* p. 221

[39] On these two roles, see Janowitz, *op. cit.,* pp. 46-51. On the principle of rotation, see e.g. C. Wright Mills, *The Power Elite,* New York, 1956: Oxford University Press, p. 194.

awards, directing the members' attention to the legitimized ways and means of making a career.[40]

Among the professions, corporateness may also be enhanced because of the fact that the members of the profession often are recruited from similar social strata. This point is somewhat less relevant to the military profession since empirical data show that officers in many countries are fairly broadly recruited (see chapter 2). Altogether however, it appears to be a reasonable hypothesis that if the various professions are compared with regard to the members' sense of group solidarity, the military profession would rank among the absolute top.

[40] See Bernard Barber, *op, cit.*, p. 19

Summary and conclusions

1. This chapter has investigated the three main elements in military professionalization$_2$, namely (a) theory, (b) ethics, and (c) corporateness.

2. A professional *theory* consists of a set of doctrines that the members perceive as fundamental to their function and practice; in the military, the doctrines of strategy, tactics, and logistics. The theory is a basis for professional autonomy and prestige. Professional schools, journals, symposia, and other media of communication are employed to transmit the doctrines to new generations of members.

3. The behavior of the professional vis-a-vis his client(s), his colleagues, and the public is regulated by a number of *ethical rules*. These rules fulfill the double function of protecting others against the abuse of professional monopoly, and defining the professional as a responsible and trustworthy expert in the service of his client(s).

3.1 An important potential source of strain in the relationship between the military profession and its client, the state, is the fact that the state is both client and superior at the same time: in disputes, however, the role of superior is assumed to prevail over that of client, i.e., the state and not the military profession is supposed to be the best judge of the soundness of military advice.

4. The *corporate character* of the military profession stems from the long period of formal education; codes of conduct, rituals, and symbols; the existence of specialized media of communication; the system of rotation between staff and command functions; and professional rewards directing the members' attention to legitimate ways and means of social ascent.

Professional corporateness has come to replace the earlier basis of cohesion on the basis of common class characteristics. Since social recruitment is now in many countries so diffuse as to exclude the possibility of unified class perspectives, intra-professional manipulation tends to take over the function of instilling values and outlooks common to the members. The degree to which the military acts subordinately as an "instrument of the state" will be affected by the impact of military professionalization (p_2) and its effectiveness in bringing about indigenuous outlooks and corporate interests. Civilian control will be impaired to the extent that civilian outlooks and interests conflict with those of the military. This calls for a discussion of what the particular military outlooks and interests may be; and that discussion will be the subject of the next few chapters.

FOUR
OCCUPATIONAL VALUES AND THE MILITARY MIND

In this chapter I will try to identify some main elements in the oft-discussed concept of "military mind". First, however, some more general arguments have to be examined. Is there a military mind? If so, what mechanisms give rise to it? Is it possible to state a definition of "professional minds" in general? Although these problems have a long tradition in sociology, from Herbert Spencer and Emile Durkheim onwards, I will concentrate on more contemporary arguments.

1. Is there a military mind?

Controversy over the existence of what might be referred to as the military mind has been a persistent feature in debates, public and scholarly, on the military establishment. Some have outrightly denied that military men are characterized by values and outlooks that differ from those of "civilians"; as often as not, one has tended to forget that "civilians" cannot be treated as one homogeneous group. For patriotic and allegedly democratic reasons, the "there's really no difference" position has been one official line of defense by supporters of the military against those critics who have pointed to the potential impact of military values on society. For instance, a couple of years after the end of World War II, US Secretary of War Robert P. Patterson stated that.

> I give it as my experience that there is no set type of military mind. There are marked characteristics. There is a highly developed sense of duty, a standard of behavior that is stricter than the average. There is a military method, a way of doing business. But I have never seen the signs of a military mind that could be identified as a single type, any more than there is a lawyer's mind, an engineer's mind or a merchant's mind. Mental equipment and outlooks on life vary as much in the Army and Navy as with other occupations and callings.[1]

In a similar manner, at the height of the Korean War, a *Fortune* article argued that it is "somewhat difficult" to "class the military as distinct from the American mind — as difficult as it would be to point to the business mind as a unique and separate phenomenon".[2]

In the democratic state, the ideal-typical image of the military man is the citizen soldier, arising from the people, fighting for the people and the existence of the state. In this picture, there is little room for the

---

[1] Address before the Alumni Association of Columbia University, June 3, 1947. Reprinted in *Infantry Journal*, vol. 61, July 1947, p. 13.

[2] *Fortune*, February 1952, pp. 91 ff.

acknowledgement that the very role of professional soldier may involve values, attitudes, and outlooks that, at times, may be sharply at variance with the political preferences and perceptions among large groups of the general public. Especially in times of crisis, political and military leaders prefer the picture of the army as part of the *Volksgeist:* and crisis or not, to the military men themselves, the idea that they may be in any important sense different from those they have chosen to defend is usually unpleasant. [3]

However, even among military men one may encounter statements conceding to the effect of professional influences, and modifying the "there's no difference" position into a sociologically more valid picture. The following quotation is from an article by a U.S. Navy Commander:

> There is a military mind if there is a judicial mind, an ecclesiastical mind, a legal mind, a medical mind, and so forth. In short, a military mind exists if we admit a professional training on subjects peculiar to the military field as a basis for service to society. [4]

The notion of a singular and specific military mind is here broadened to the more general hypothesis of professional values as the products of intra-occupational socialization. This view, of course, is similar to analyses by such sociologists as Durkheim, Mannheim, Laski, and Merton. [5]

Laski notes the traditional conservative attitudes of lawyers, and suggests as a partial explanation that the lawyer's habits "are rooted in precedence and tradition. He is seeking to predict for his clients the expectations of to-morrow in terms of past certainties. His effort is therefore obviously directed towards the approval rather of stability than change". [6] The day-to-day preoccupations of the legal expert thus forms a basis for a specific *Weltanschauung*. In like manner, Merton has lined out the impact of organizational rules and requirements in bringing about a certain bureaucratic personality structure. [7] And in a more general approach, Mannheim has analyzed the social bases of political thought and points out how the social situation of the individual determines his interests, and how these in turn determine his conception of the social process. [8]

A study by Rosenberg gives considerable empirical support to the idea that persons aspiring to different occupations tend to differ from one another in terms of certain values and attitudes, such as "orientation to people", "extrinsic value-orientation", and so on. [9] Thus, besides the possible effects of intra-professional socialization, these anticipatory processes also contribute to the homogenization of occupational "minds". And, as we shall see shortly such "minds" are also in part the products of selection mechanisms.

2. Four processes of homogenization

In the article on the social recruitment of the Norwegian officer corps quoted above, [10] Francisco Kjellberg points out that the broadened social representativeness does not seem to have had much effect on the

[3] Compare the statements by Bradley and Göransson, ch. 5, footnotes 9 and 10.

[4] H.E. Smith, "What is the Military Mind?", *U.S. Naval Institute Proceedings,* vol. 79, May, 1953.

[5] Emile Durkheim, *Professional Ethics and Civic Morals,* New York, 1958: Free Press. Karl Mannheim, *Ideology and Utopia,* N.Y., 1936: Harcourt, Brace, and World. Harold Laski, *Democracy in Crisis,* Chapel Hill, 1933: University of North Carolina Press. Robert K. Merton, *Social Theory and Social Structure,* Clencoe Illinois, 1957: Free Press

[6] Laski, *op.cit.,* p. 141–143

[7] Merton, *op.cit.* ch. 6

[8] Mannheim, *op.cit.,* p. 163.

[9] Morris Rosenberg, *Occupations and Values,* Glencoe, Illinois, 1957: Free Press.

[10] Ch. 2, footnote 11.

professional outlooks of the military: "the group now derives its main characteristics from a particular culture, developing from occupational activities and quite irrespective of social composition".

But "occupational activities" only represent one factor among several that work together in bringing about professional outlooks. A full representation of such mechanisms will have to take into account at least the following four areas.

(1) The specific *interest or motivation for a certain vocation* among potential candidates functions as an important filter in occupational choice. As Rosenberg points out with regard to the USA, theoretically the American student has about 40.000 occupations from which to choose, but in reality only a small part of this broad spectrum is screened. Some occupations are *not socially appropriate* given the social status of the individual; some are *not possible*, because of educational or other limitations; finally, some occupations are *not desirable* given the individual's values, attitudes, and personality characteristics.[11] The discussion below deals primarily with the third of these issues.

Each occupation has a set of characteristics, which in more or less distorted forms are transmitted to the potential recruits, who then try to match their values with their images of the occupations. These impressions may be construed as being the products of (a) direct communication between the potential recruit and persons working within the various occupations, (b) indirect communication with occupational members through the mediation of others, and (c) published material about the occupation (recruitment pamphlets, articles in the press, television and radio programs, etc). Through one or more of these channels, the aspirant is exposed to the goals of the occupation, i.e. he acquires information about the tasks he will have to perform. To a certain extent, therefore, the desirability of an occupation will be dependent upon the individual's perceptions of its goals. As Rosenberg shows, persons who want to "work with people rather than things" tend to choose "people-oriented" occupations like social work and teaching, while others with less strong preferences in this area favor occupations such as business and finance, or sales promotion.[12]

Thus among the applicants to a profession there will be an over-representation of individuals who are attracted to its visible characteristics and/or what they perceive as being its major advantages. This is amply verified by empirical research: persons who want to give vent to their (real or imagined) creative abilities try to enter, for instance, journalism or the arts; if the dominating motive is to earn money, their choices tend towards business, etc.[13] Potential candidates to the military profession emphasize the prospects of outdoor life, good comradeship, and patriotism.[14] To the extent that vocational expectations coincide with a specific set of values, therefore, this process of self-selection tends to contribute to professional value-homogenization.[15]

(2) Many professions utilize particular *screening procedures* in order to eliminate candidates of less than required intellectual and physical

[11] Rosenberg, *op.cit.*, p. 4

[12] *Ibid.*, pp. 26 ff

[13] *Ibid.*, loc. cit.

[14] Robert Erikson, *Yrkesval och officersrekrytering*, MPI report no. 31, Stockholm. 1964: Militärpsykologiska institutet (mimeo)

[15] Usually there will also be a number of persons who are indifferent to the occupation and who make their choice more or less at random. Conceivably, the probability that they will after a certain period quit their jobs for more suitable alternatives is greater than for persons who were more committed initially.

ability. For young men aspiring to enter the military, this process typically involves the taking of an IQ test and a medical examination. Often, however, the applicant's ideological viewpoints are also examined; for instance, in Sweden one of the requirements to be fulfilled when applying for officer training is having "citizen spirit" — a general formulation, at present so interpreted by military officials as to exclude candidates with known Communist or Nazi sympathies. Thus, there is a deliberate selection of candidates with what is officially regarded as congenial — or, perhaps more correct, not uncongenial — attitudes.[16]

But the number of applicants turned down for ideological reasons by the military authorities, at least in Sweden, is usually small, since self-selective processes tend to strongly narrow down the size of the recruitment base in favor of the congenial candidates.

(3) Proceeding to the internal conditions of the profession, there is also a *continuous selection of people with the "right" attitudes within the profession*, since the chances for promotion are not equal but tend to favor candidates who conform to their superiors' expectations. This is illustrated by a well-known panel study in *The American Soldier*.

The attitudes of a sample of 378 privates were surveyed in September 1943. Their attained rank as of January 1, 1944, was ascertained; a fifth were found to have become privates first class (PFC's), and four fifths were still privates. The following table lists the percentages of promotion among privates who did and who did not accept the soldier role. (The answer "soldier" to the question "If it were up to you to choose, do you think you could do more for your country as a soldier or as a worker in a war job?" defines acceptance of the soldier role).

TABLE 5

	Accepted the soldier role	Did not accept the soldier role
High School graduates 25 and over	45 (20)	42 (50)
High School graduates under 25	29 (45)	24 (41)
Others 25 and over	25 (16)	14 (78)
Others under 25	14 (49)	8 (79)

Source: S.A. Stouffer et al., *The American Soldier. Adjustment during Army Life* (Princeton, 1949), p. 150.

From these figures it is clear that men, within a given subgroup by age and education, who had congenial attitudes also had the better chances of promotion. The findings were replicated on other samples of enlisted men and officers. The authors conclude: "The data ... support the expectation that within the Army, as in perhaps any institution and especially any authoritarian institution, the price of advancement was at least a minimum conformity with the system, in mind as well as in action."[17]

[16] There are also examples of ethnic and tribal recruitment principles. The colonial powers (Great Britain, France, and the Netherlands) in setting up overseas armies often tried to recruit enlisted personnel and officers from tribal groups remote from the central capital, from minority groups, and from groups with limited independence aspirations. The recruitment of personnel in Nigeria is stratified according to regional factors: 50 per cent of all recruits are taken from the North, and the rest in equal proportions from the eastern and western regions. (Morris Janowitz, *The Military in the Political Development of New Nations*, Chicago, Ill. 1964: University of Chicago Press, pp. 52-54.)

[17] Stouffer et al., *op.cit.*, pp. 263-264

It should be noted that the figures in table 5 represent wartime conditions, during which the pressure for rapid promotion of personnel is stronger, and the criterion of "acceptance of the soldier role" probably has less relative weight than it would have during peacetime. Had the survey been carried out after the war it is probable that the differences in promotion rate between the two categories would have been more accentuated.

(4) Finally, the effects of these selection processess are further amplified by *professional socialization and training*. The military is a good example of Huntington's general description of the professional milieu.

People who act the same way over a long period of time tend to develop distinctive and persistent habits of thought. Their unique relation to the world gives them a unique perspective on the world and leads them to rationalize their behavior and role. This is particularly true where the role is a professional one. A profession is more narrowly defined, more intensely and exclusively pursued, and more clearly isolated from human activity than are most occupations. The continuing objective performance of the professional function gives rise to a continuing professional *weltanschauung* or professional "mind".[18]

As we shall see later, a variety of research findings point to the existence of a particular military culture. For example the military elite tends to be clearly more conservative than the population at large and even in comparison with other professional groups.[19] and among young men who seek to become officers, persons with authoritarian attitudes are over-represented compared to those who enter other occupations.[20]

Unfortunately, however, the research designs usually employed make it difficult to assess the impact of this culture, and of professional socialization and training, compared to selection at induction and during the career. Sociological studies of the military profession have been carried out with cross-sectional designs; hence, differences in opinion (e.g., political preferences) between various ranks or age groups cannot with any certainty be attributed to indoctrination processes, as against selection.

To sum up then, what we can say is that value-homogenization takes place *both* through selection and indoctrination, and in spite of common assumptions, perhaps not even primarily through the later. To a large extent, people who seek entrance into the military are already a rather specific group, since many have matched their career expectations with their perception of the military profession; and under the process of approaching military life they probably adjust still further to their future environment (i.e., through anticipatory socialization).[21] On the whole, we have to leave the question of the relative importance of the above-mentioned four processes for further research.

3. The professional mind: a definition

The professional mind is a product of selective *and* indoctrination processes. Our definition therefore has to be one which is equally compatible with both, and we have to avoid postulating that it is only (or even mainly) brought about by socialization after induction. A minimum

---

[18] Huntington, S.P., *The Soldier and the State*, Cambridge, Mrs., 1957: Harvard University Press, p. 61. Anatol Rapoport has in a general exposition of "the blindness of involvement" pointed out some dysfunctions of know-how and expertise: "In short, the interest in an activity, its challenge, the feeling of mastery it imparts to its successful practitioners usually far outweigh both the acquisitive impulses and the moral considerations associated with it. Preoccupation with craftmanship dulls moral sensitivity, just as preoccupation with large-scale affairs dulls empathy." *Fights, Games, and Debates*, The University of Michigan Press, 1960, p. 272

[19] See below, ch. 10

[20] See below, ch. 8

[21] Rosenberg, *op.cit.*

definition is that the professional mind is a set of relevant attitudes U,V,W,....Z, on which members of the profession have, on the average, more extreme (or higher, if we standardize all variables in the positive direction) positions than members of any other identifiable occupational group, or of the total population; all contaminating variables are assumed to be held constant.[22]

Note 1. One or a few of U,V,W,....Z may be assumed to relate more closely than the others to the goals of the profession or, in other words, to be of greater functional significance to the goals than the rest of the variables. Hence it (they) will be more dominant and important for adjustment to the professional role.

Note 2. For increasing levels of professionalization, we expect the professional mind to become more "pure" or "crystallized"; thus the higher the level of professionalization, the more often we expect to find individuals with combinations of extreme (or high) values on U,V,W,....Z. This may take place through selection, or indoctrination, or both. Through *selection* individuals with "crystallized" sets of attitudes may by favored before others; through *indoctrination* individuals with initially less "crystallized" attitude sets may be influenced in the direction of greater "crystallization". As mentioned above, previous research does not allow conclusions as to which process is the more important.

There remains the question why I have chosen certain variables (U,V,W,....Z) and not others (M,N,O,....T) as being especially "relevant". The answer, to be dealt with in a more detailed way below, is the following. First, previous analyses of the military profession give a certain guidance: as we shall see, there is more intersubjective agreement in this area than might have been expected, taking into consideration the wide differences among writers in their sentiments toward the military. Second, the relevance of a certain variable in the military mind can at least partly be derived from the goals of the profession. We will turn to these issues in the next two sections.

[22] For instance, in investigating authoritarianism we will have to control for differences in education and intelligence between the profession and the groups with which we compare it. It will always be possible, of course, to compose by deliberate selection a sample of individuals exhibiting the attitudinal characteristics even more strongly than the members of the profession. From what has been said, however, it follows that such a group will not be statistically representative of a particular *occupational* group nor of the total *population.*

4. Components of the military mind

The debate on the military mind has more demonstrated differences between those who advocate that the military mind is a myth and those who concede to its existence, rather than different opinions as to the content of it, once its existence is admitted. Compare, for instance, the following four quotations, the first two by authors who are essentially critical of the military establishment. (C. Wright Mills and John M. Swomley, Jr.), the others by observers who are mainly sympathetic (Samuel P. Huntington and Charles O. Lerche).

Mills says:

Internally, to the extent that the whole system of lifetraining has been successful, they are also reliably similar in reaction and in outlook. They have, it is said, "the military mind", which is no idle phrase; it points to the product of a specialized bureaucratic training; it points to the results of a system of formal selection and common experiences and friendships and activities — all enclosed within similar routines. It also points to the fact of discipline — which means instant and stereotyped obedience within the chain of command. The military mind also indicates the sharing of a common outlook, the basis of which is the metaphysical definition of reality as essentially military reality. Even within the military realm, this mind distrusts "theorists", if only because they tend to be different: bureaucratic thinking is orderly and concrete thinking.[23]

Similarly, Swomley argues:

Military interests involving the achieving or maintaining of a large military establishment make it necessary . . . to build fear or hatred of an actual or potential enemy. . . .Military interests tend to view foreign policy in essentially military terms. They are unwilling to have civilian leaders take a larger picture into account. . . .Sometimes the problem /in civil-military relations/ is simply military impatience with the slower processes of democratic discussion and compromise.[24]

Huntington summarizes the military ethic in the following way:

The military ethic emphasizes the permanence, irrationality, weakness, and evil in human nature. It stresses the supremacy of society over the individual and the importance of order, hierarchy, and division of function. It stresses the continuity and value of history. It accepts the nation state as the highest form of political organization and recognizes the continuing likelihood of wars among nation states. It emphasizes the importance of power in international relations and warns of the dangers to state security. It holds that the security of the state depends upon the creation and maintenance of strong military forces.[25]

Finally, Lerche in an official U.S. Army handbook describes the military outlook in the following terms:

One basic proposition should be made at the outset: there is a generic difference between the way military personnel approach and solve foreign policy problems and the way their civilian opposite numbers do the same thing. . . .Probably the most significant single characteristic of the military's approach to foreign policy is a strong belief in "can do". There is a great temptation among orthodox policy-makers, when complex and ambiguous situations are faced, to delay commitment and action until only one course becomes feasible. . . .To this tendency military spokesmen generally find themselves opposed; the American military tradition strongly emphasizes the necessity of solving problems rather than merely enduring their consequences. . . .Charged professionally with a conscious devotion to the United States as a national state, officers have fewer inhibitions in their way when they seek to place national interests above personal, group, regional, or party concerns. . . .In any discussion of national policy, the military are more likely to follow a direct and single line focusing on the national interest than any other identifiable group.[26]

The authors differ in the degree to which they select negatively loaded terms in their descriptions of the military mind, but agree in several essential respects: the military mind is *nationalistic* (Swomley: "build fear or hatred of an actual or potential enemy"; Lerche: "focusing on the national interst"), *authoritarian* (Mills: "instant and stereotyped obedience"; Swomley: "they do not like to have their decisions questioned"; Huntington: "stresses the importance of order, hierarchy . . ."), and *alarmist*[27] Swomley: "tend to view foreign policy in essentially military terms"; Huntington: "recognizes the continuing likelihood of wars"). These similarities among writers fundamentally different in general orientation tend to support the concept of military mind as being reasonably free from value-distortions.

As for the *pessimistic beliefs on human nature*, similarities among authors are less easy to demonstrate, since this factor has not been extensively commented upon, with the exception of Huntington's *The*

[23] C. Wright Mills, *The Power Elite*, New York, 1956: Oxford University Press p. 195

[24] John M. Swomley, Jr., *The Military Establishment*, Boston, 1964: Beacon Press, pp. 250 ff.

[25] Samuel P. Huntington, *The soldier and the State*, p. 79

[26] Charles O. Lerche, Jr., "The Professional Officer and Foreign Policy", *Strategic Subjects Handbook*, U.S. Army Command and General Staff College, Fort Leavenworth, Kansas, September 1967, R 1800-1, p. Li-5 f.

[27] For a penetrating historical discussion of alarmism, see Alfred Vagts, *A History of Militarism*, New York, 1959, Hollis and Carter, ch. 11.

Soldier and the State. It is included here for two reasons: first, basic orientations of this kind have been demonstrated to be important predictors of occupational choice;[28] second, this attitude is clearly in balance with the goals of the military profession, and to the extent that philosophical images of man at all are prevalent among the military, they are likely to stress tendencies of violence, irrationality, and weakness, rather than their opposites. The discussion on this point will be mainly theoretical and speculative: few data are available, and only in a very general way can the relationship between the variable and the others be demonstrated.

Finally, one important variable in the military mind is *political conservatism*. This hypothesis is entered here on the basis of a large body of theoretical and empirical studies[29] and does not need further elaboration in this context. The data will be discussed below, ch. 10.

5. The military mind: a summary sketch

The primary goal of the military establishment is the security of the state, the protection of which is the major preoccupation of the military profession. We may expect aspirants and members of the profession to hold attitudes compatible with this goal.

On the basis of the previous section, the U,V,W,...Z of the military mind may be defined as (1) nationalism, (2) pessimistic beliefs on human nature,[30] (3) alarmism (i.e., "pessimistic" estimates on the probability of war), (4) political conservatism, and (5) authoritarianism. It seems a reasonable hypothesis that (1), (2) and (3) are deriving from the specific military goals; therefore we would expect them to be *independent of internal structural changes of the military organization*. Changing the forms of organization, for instance in the direction of more manipulative rather than dominant leadership, will probably not bring about changes in this basic subset of attitudes; neither do we have any reason to expect such changes if the social recruitment of the military is broadened. Thus, only revision of the basic goals of the profession, for instance from being a force for territorial defense to a force for international peace-keeping would cause changes in the fundamental subset (primarily in nationalism, since the use of military force still implies both (2) and (3)).

In the Western countries, nationalism has tended to favor status quo, and nationalist policies as well as patriotic sentiments have been supported primarily by conservative civilian factions and parties. Hence, in these countries (1) and (4) are in basic agreement, and to the military man conservatism and nationalism represent highly compatible elements. However, in a number of Third World countries this proposition is less valied (e.g., Egypt, Algeria, Peru under Velasco, and Bolivia under Torres): here we find at least occasionally that the nationalist inclination of the military does not exclude even relatively egalitarian and pro-leftist sentiments. It seems reasonable, therefore, to view the political outlooks (4) of the military as depending on nationalism (1) rather than the other

---

[28] Rosenberg, *op.cit.*,

[29] See Huntington, *op.cit.*; Morris Janowitz, *The Professional Soldier*, Glencoe, Illinois, 1960: Free Press ch. 12; Eric Waldman, *The Goose-Step is Verboten*, New York, 1964: Free Press ch. 9; Walter Korpi, *Social Pressures and Attitudes in Military Training*, Stockholm, 1964: Almqvist o Wiksell pp. 82-95; Alfred Vagts, *A History of Militarism*.

[30] This professional bias is probably best illustrated by handbooks in counterespionage, such as *Maskerad front (Masked Front)* published by the Swedish defense. The manual stresses the permanent and immediate threats to each citizen. "The front with the sharp bullets, the bombs, and the shells – that is the 'outer front'. What you have read about here is the 'second front' – with the silent weapons – 'the masked one'. On that front we are all attacked – soldier or not soldier. On that front there is never peace – there we fight at this very moment – even you. Against you stand the soldiers of the masked front – the subversives to be – *the enemies of your country.*" *Maskerad front*, Handbok i säkerhetstjänst, Stockholm 1964: Försvarets Bok- och Blankettförråd, p. 5)

way around; in other words, when the military man makes his political choice he tends to favor parties or factions giving nationalist policies high priority.

Finally, (5) authoritarianism, The specific organizational forms of the military profession, with its emphasis on efficiency, coordination, and subordination, can be assumed to put a premium on authoritarian rather than democratic attitudes. We will examine the often made assumption that the military, because its organizational setting is authoritarian, will exhibit attitudes that are compatible with this setting. Authoritarianism, therefore, is not primarely dependent on the professional *goals*, but rather on military organizational characteristics. Hence in this case we *would* expect authoritarian attitudes to change with, for instance, a transition from dominative to manipulative forms of command, and to be less prevalent among military men in managerial as compared to command positions.

In the following chapters, we will turn to some data and some further arguments to support the inclusion of (1) to (5) in the set of variables that make up the military mind.

FIVE

NATIONALISM: AXIOM OF MILITARY IDEOLOGY

The overriding concern of the military profession is centered around the security of the state. With few exceptions – primarily the supranational forces of the United Nations – national defense against actual or potential enemies forms a primary reason for the existence of military forces, large or small. "All members of a society have an interest in its security; the state has a direct concern for the achievement of this along with other social values; but the officer corps alone is responsible for military security to the exclusion of all other ends".[1]

Johnson writes about the emergence of nationalism as a political force in Latin America around the turn of the nineteenth century, and points out its impact on the military as an evolving institution.

> The disputes that arose from the efforts of the republics to guarantee or extend their boundaries sprang from the influence of the new nationalism; and although most of them were settled by arbitration, peaceful negotiations broke down or threatened to break down from time to time. With the examples before them of Chile, which made war both glorious and profitable by seizing valuable mining territories from Bolivia and Peru, and of Brazil, which used its fighting forces to back up diplomatic efforts that led to the acquisition of territory at the expense of weak neighbors, it was easy for politicians to justify expanded armies and navies in the interests of national sovereignty. Nationalism thus became the ideology on which the military grew By the end of the century the armed forces had become the agencies for carrying out the spirited international policies of the republics or, alternatively, symbolizing the defense of the national sovereignty.[2]

Today, the military in Latin America are "superpatriots", according to Johnson's description.[3]

Talcott Parsons has called attention to the central position occupied by nationalism in his essay on patterns of aggression in the Western world. The nationalistic attitude is partly supported by the social fact of the organization of our civilization into nation-states, in which allegiance to one's government has become the ultimate loyalty, "the one which could claim any sacrifice no matter how great if need be". Nationalism is an inherent element in the syndrome of fundamentalist sentiments, and forms an "ultimate test of altruism and sincerity", characteristics which opponents of the fundamentalist doctrines often are accused of lacking.[4] The fundamentalist outlook has a close resemblance to military absolutist doctrine, as described by Janowitz, and according to which U.S. long-term political goals are total supremacy, supported by the

[1] Huntington, *The Soldier and the State*, Cambridge, Mass., 1957: Harvard University Press p. 15.

[2] Johnson, *The Military and Society in Latin America*, Stanford, 1964: Stanford University Press p. 68-69

[3] *Ibid.*, p.251

[4] Talcott Parsons, "Certain Primary Sources and Patterns of Aggression in the Social Structure of the Western World", in *Essays in Sociological Theory*, New York, 1964: Free Press.

"fortress America" strategy.[5] This doctrine has had its foremost representatives among the leaders of the Strategic Air Command. For instance, consider the following excerpt from *Design for Survival* by General Thomas S. Power, former Commander in Chief of the SAC.

[5] Janowitz, *The Professional Soldier,* Glencoe, Illinois, 1960: Free Press, ch. 13, esp pp. 273-277

> During my over seven years as Commander in Chief of the Stategic Air Command I talked to scores of diverse groups of leading citizens who visited SAC Headquarters because they had a deep and sincere interest in getting firsthand information on the Free World's most powerful military deterrent to aggression. Virtually all of these people – professional men and women, ministers, labor leaders, industrialists, businessmen – voiced their full support for maintaining a strong and credible military deterrent as the principal safeguard of an "honorable peace." But there was one question that came up quite frequently: For how long do we have to keep it up? The only answer my conscience permitted me to give was, "For as long as our survival is threatened; that is, indefinitely."
>
> Occasionally the subsequent discussion would lead to another question, namely, whether the time would ever come when our survival might no longer be threatened and we could safely discard our weapons. This, in turn, would bring up the subject of one-world government as a means for ending the arms race, and I would find that quite a few of my listeners believed this to be a worthwhile goal, however difficult and distant it might be. But when I would question them regarding the implications and dangers of the one-world principle, it became patently clear that they had given little if any thought to it.
>
> It is significant that these and many other people who firmly support a national policy of deterrence based on superior military strength are equally convinced or, at least hopeful, that one-world government offers the ultimate solution for the problem of survival in the nuclear age. Personally, I fail to understand how a person can profess to be willing to give his all for national survival and at the same time advocate a goal which, in effect, means surrender of our national sovereignty and integrity. Still, there is widespread belief in both the desirability and feasibility of one-world government and, indeed, much indication that we are going down that road.[6]

[6] Thomas S. Power, *Design for Survival,* New York, 1965; Pocket Books Inc., pp. 70-71

Interestingly, the main argument is not that a world government would be *ineffective*, but that peace treaties and international agreements, possibly leading to a supranational government, are *dangerous,* because national sovereignty would have to be abandoned. "National sovereignty and integrity" is the first principle of the absolutist doctrine.

The obvious success of the military establishment to perpetuate its own existence even during periods of relaxed international tension is partly explainable by its ability to evoke fundamentalist sentiment, and by the reluctance of politicians to question military appropriations for fear of appearing unpatriotic and disloyal.

On the other hand, when, at times, a more internationalistic outlook prevails in society, the military man often resents being described as a nationalist. Janowitz points out how tendencies toward ethnocentrism are deliberately held back in the education of American officers in order to eliminate the civilian contempt of the military mind, an image which is unfavorable to the prestige of the military profession.[7] But, at any given period, whether the dominant mood among the public is nationalistic or internationalistic, the security goal cannot be implemented very effectively by people who are indifferent to it; just as people who are indifferent to the practice of healing within the medical profession do not become very good doctors. Although the public relations of the military establishment sometimes require the dissemination of

[7] Janowitz. *op.cit.,* pp. 12-13

the image of the officer as non-ethnocentric, the basic value of the profession is nationalism; and as long as the means of security policy are primarily those of territorial defense, the military will constitute a highly influential body in support of these means, and keep exerting heavy pressure on the economy in order to increase — or at least to hinder the reduction of — defense appropriations.

While civilian advocates of social reform often see the military as one major obstacle to a restructuring of the economy, this view is resented by military officers, since they perceive national goals as determined essentially by civilians and their own role as impartial and neutral servants of the state. This puristic "animated machine" view of the military establishment[8] is exhibited, for instance, in an article by General Omar N. Bradley. He argues that "the projected outlay for defense is not in reality a military budget; it is a civilian budget".

> Civilians are in charge. In the Defense Department, the budget is controlled and finally approved, not by the Joint Chiefs of Staff but by the four civilian secretaries. ... Economically, politically and military, the control of our country resides with the civilian executives and legislative agencies, and thus ultimately in the hand of the voters and organized civilian groups interested in good government.[9]

The same type of argument is used also by representatives of the Swedish military. An Army general says:

> The political direction of the Defense is carried out by the Government, which thus has the "power" over the Defense. /The Commander-in-Chief is/ in relation to the Government and the Parliament /Riksdagen/ an authority with informative, proposing and executive functions. The executive jurisdiction of the Commander-in-Chief is regulated by statutes, Royal Charters and general orders. Laws and appropriations are decided on by the Parliament.... /T/he head of the Government — under the King — stands out by virtue of his office as the national leader of defense preparations in peace and for the war effort in general.[10]

This way of clarifying civil-military relationships does not adequately take into account that in actual practice the power of expertise — and the disinterestedness in military affairs among many politicians — will tend to compromise the effectiveness of constitutional checks and balances; nor does it suffiently emphasize the important fact that the military through pressure-group tactics and lobbying have substantial possibilities of co-opting politicians.[11] But in the current debate on the military system, the fears voiced are not primarily that the military deliberately upset constitutional rules, but rather that they *within the framework of those rules* have acquired considerable political and economic power.[12] Some problems related to this issue will be treated at some length below (ch. 11).

With regard to empirical data, there have to my knowledge been no research results published on nationalism among the military, which could be of direct relevance to the above discussion. However, some indirect evidence can be cited. Waldman has conducted a survey among soldiers of different ranks in the West German Bundeswehr, and also has comparable data on a random sample of the total population.[13] While his general interest was the evaluation of the program of *Innere Führung*

[8] For the origin of this term, see below, ch. 13, footnote 18,

[9] "Should We Fear the Military?" *Look,* vol. 16, March 11, 1952, p. 35

[10] Curt Göransson, in *Fred och försvar under 60-talet,* Stockholm, 1961: Esselte

[11] For a general treatment of the problems of relations between experts and politicians, see Piet Thoenes, *The Elite in the Welfare State,* New York 1966: Free Press, esp. chs. 5 and 9; also Don K. Price, *The Scientific Estate,* Oxford University Press, 1965.
Regarding military pressure group tactics, see Janowitz, *The Professional Soldier,* chs. 17-19; C. Wright Mills, *The Power Elite,* New York, 1956: Oxford University Press pp. 199-202; "American Militarism", *Look,* vol. 33, Nos. 16 and 17, 1969

[12] See, for example, John Kenneth Galbraith, *How to control the military* (Swedish edition *Att hålla Pentagon i schack,* Stockholm, 1970: Tema)

[13] Eric Waldman, *The Goose-Step is Verboten,* New York, Free Press, 1964, ch. 9

and the attitudes of the military towards NATO, the response alternatives to at least one of his questionnaire items could be used as indices of a nationalistic inclination. The respondents were asked: "What is the most important contribution of the Federal Republic to NATO?" and were presented the following alternatives:

Answers (a) and (b): "That it provides territory for the NATO military forces", and "that it took over a large portion of the economic and financial burden".

Answers (c) and (d): "That it provides a significant military contribution", and "that it provides its share to the prevention of Soviet aggression".

The second two alternatives reflect an attitude more hostile to the "out-group" than the first two. Hence we would expect the military sample more often than the civilian respondents to select the latter alternatives, which is indeed the case.

TABLE 6

	Per cent		
	Answers (a) and (b)	Answers (c) and (d)	No information
Total population	36.9	47.5	15.3
Bundeswehr:			
military ranks			
Recr.	15.9	78.7	5.1
Cpl.	10.9	85.7	3.1
Sgt	6.1	92.4	1.3
S-NCO	4.1	92.4	3.2
OC	5.7	93.6	0.5
Off (Lt. Capt.)	4.0	95.8	—

Source: Eric Waldman, *The Goose-steep is Verboten*, p. 238. *Legend:* Recr = recruit; cpl = corporal; sgt = sergeant; S-NCO = senior non-commissioned officer (Feldwebel); OC = officer candidate; lt = lieutenant; capt = captain.

Only part of the original table is reproduced here, but the other statistical breakdowns presented by Waldman give essentially the same picture.

The proportion of agreements to alternatives (c) and (d) is not only markedly larger among the military than among the total population, but it also tends to increase with rank, which is expected because of the functional role played by this attitude in military professionalization.

It should be emphasized that nationalistic attitudes may be perfectly compatible with positive attitudes to the role of other countries in a common military bloc, since the combined efforts are assumed to strengthen the position of each individual country against the common adversary. Thus in this case the concern for the security of the state is transferred into a favorable opinion of the defense community. For instance, Waldman shows that the military in the Bundeswehr are more inclined towards a positive attitude to NATO than the general population, which seems to be largely indifferent.[14]

A study by Eva-Stina Bengtsson among Swedish reserve officers at the University of Lund shows that the officers, more often than comparable civilian students, tend to endorse the item "to obtain peace, countries will have to keep national armies". As expected, the officers also markedly differ from their civilian fellow students in their attitudes towards Swedish disarmament: to the items "Sweden should disarm only if all other countries disarm", or "Sweden should never disarm", 87 per cent of the reserve officers agreed, as against 50 per cent of the civilian students. And while almost half of the civilian sample agreed with the statements that "Sweden should disarm even if no other countries disarm", or "Sweden should disarm if our neighbor countries disarm", only 8 per cent of the reserve officers answered in the affirmative on these items.[15]

Although the available empirical material is scant, it tends to support rather than disqualify the assumptions developed above. As the Bundeswehr data show, the military tend to deviate markedly from the general population with regard to out-group hostility. And as the Swedish evidence demonstrates – although hardly surprising – the military role is associated with disapproval of arms reduction. Such disapproval, of course, is a natural counterpart to the military man's concern for the security of the nation-state, and to his attachment to the traditional means of territorial defense to further this goal.

[14] *Ibid.*, p. 234-235

[15] Eva-Stina Bengtsson, "Some Political Perspectives of Academic Reserve Officers", *Journal of Peace Research,* no. 3, 1968, pp. 293-305. The respondents are not statistically representative of the total population of Swedish reserve officers. The differences reported remain when political party preferences are held constant.

SIX
IMAGES OF HUMAN NATURE

On the basis of psychological balance theory, we expect a relationship between the type of tasks with which the professional is preoccupied, and the type of rationale by which he motivates his particular activity. As the results from Rosenberg's study cited above imply, successful professional adaptation demands that the individual's values and philosophy of man are in basic agreement with the primary goals and priorities of the profession.

The defense establishment is preoccupied with the rational planning and, if need be, execution of military operations for the protection of the nation against the aggression of others. National sovereignty is one fundamental element in the military man's beliefs; the image of "the others" as basically untrustworthy is another. Conversely, internationalism and belief in the trustworthiness of other nations will be dissonant with the military role. Or, in the words of Samuel P Huntington:

> The existence of the military profession presupposes conflicting human interests and the use of violence to further those interests. Consequently, the military ethic views conflict as a universal pattern throughout nature and sees violence rooted in the permanent biological and psychological nature of men. As between the good and evil in man, the military ethic emphasizes the evil.[1]

Such outlooks furnish the motivational background for the military profession even in countries which have not been at war for very long periods, such as, for example, Sweden and Chile. The previously mentioned Swedish handbook on counter-espionage (see above, ch. 4, footnote 30) points out as a typical Swedish "weakness" that "we have a certain disposition to believe that people are good, a credulity which in a critical situation may lead to serious consequences".[2] In Chile, officers emphasize the need for the armed forces because of, among other things, the alleged intentions by Peru and Bolivia to recover the territory lost to Chile in the War of the Pacific (1879–1884) and Argentina's dreams of territorial expansion.[3]

We would expect persons with pro-military orientations to show a pessimistic bias with regard to the issue of man's ability to learn to avoid war. Data from Bengtsson's study of reserve officers at the University of Lund tend to support this, although the differences between the officers and their civilian counterparts are not very large. Whereas 69 per cent of the civilan students agree to the statement that man can learn to avoid war, the proportion among the reserve officers is 51 per cent. If political

---

[1] Huntington, *The Soldier and the state*, Cambridge, Mass, 1957: Harvard University Press pp. 62-63

[2] *Maskerad front*, Handbok i säkerhetstjänst, Stockholm 1964: Försvarets Bok- och Blankettförråd p. 61.

[3] Roy A. Hansen, *Military Culture and Organizational Decline: A Study of the Chilean Army*. Unpubl, doctoral dissertation, University of California, Berkeley, 1967, p. 232

[4] Eva-Stina Bengtsson, "Some Political Perspectives of Academic Reserve Officers", *Journal of Peace Research*, 1968, 3, pp. 293-305.

[5] Vagts, *A History of Militarism*, p. 340

opinion is held constant, the differences between these two groups in the "liberal" category are larger than in the "conservative" category (63 and 47 per cent of the "liberal" civilan students and reserve officers agree, as against 53 and 48 per cent among the "conservatives"); this also reflects the relationship between conservatism and a pessimistic image of man.[4] It is possible that a comparison between civilian students and young regular officers would have yielded somewhat larger differences, since the more professionalized officers are likely to hold, on the average, still more pessimistic biases.

Such pessimistic images as those noted here may have direct implications for the functions of military war planning. Most fundamentally, war plans are based on the identification of one or several enemies, and on the assumption that he (they) some time in the future will choose to attack. Identification of possible enemies and predictions regarding their aims and objectives thus become two main justifications for long-range military planning. As Vagts says, "almost as long as bureaucracies and standing armies have existed, they have recommended themselves to their contemporaries and to posterity on the score of their prescience, as being able to plan for the future".[5] But as repeatedly pointed out in the literature, such advance planning may easily become a provocative element.[6] The planning for a *possible* war may cause or contribute to causing *actual* war; and staff planning may increase rather than reduce the probability of armed solutions of international conflict.

The military view of human nature is consistent with social-darwinistic doctrine, in which military strength is seen as functional for the survival of the fittest among nations. War, not peace, appears to military writers to be the most significant characteristic of human life.

> Life is a struggle. The end of the battle is selection. The strong must conquer, survive, and reproduce; the weak submit, perish, and disappear. So it is ordained by Nature.
> Man is no exception in falling under this law, and with him races and states, the structures that Nature and himself have reared to aid him in his struggle...."Eternal peace is a dream" says Moltke, the German Commander in the Wars of Unification, "for war is a part of God's world ordinance."[6b]
> The fundamental law governing life in this world is war. Animals, insects, fish, all wage incessant war against or with each other, so that they may feed and exist. The advent of the gift of reasoning to the higher animals did not check war, it merely accentuated its use, and has caused it to become more devilish as centuries have progressed.[7]

The mere fact that war has been a common phenomenon in the annals of mankind may seemingly verify the assumption that man is fundamentally and permanently aggressive. The opposite view, i.e. that man is basicially non-aggressive (and that wars may be the products of social rather than biological conditions and hence may be checked through means of social control) has not been particularly popular among military men, in spite of the fact that history may be taken as support of this as well; after all, for most nations periods of peace have usually been considerably longer than periods of war. The military view of mankind is pessimistic. "Man has elements of goodness, strength, and reason, but he is also evil, weak, and irrational. The man of the military ethic is essentially the man of Hobbes".[8]

[6] Robert K. Merton, "The Self-Fulfilling Prophecy", in *Social Theory and Social Structure*, Glencoe, Illinois, 1957. ; Lewis F. Richardson, "Generalized Foreign Polity", *British Journal of Psychology Monographs Supplements,* 23 (1939), Anatol Rapoport, *Fights, Games, and Debates,* University of Michigan Press, 1961, ch.1; Amitai Etzioni, *Winning Without War,* Anchor Books (Doubleday), 1965, ch. 5; Anatol Rapoport, introduction to *Clausewitz on War,* Penguin Books, 1968, p. 12. For some general expositions of the contrasting views of the nature of man, see e.g. Paul Roazen: *Freud – Political and Social Thought,* New York and Toronto, Alfred A. Knopf, ch. IV; Clyde Kluckhohn, *Mirror for Man,* New York and Toronto, Mc Graw-Hill, 1949, ch. 10; Muzafer Sherif and Carolyn W. Sherif, *Groups in Harmony and Tension,* New York, Octagon Books, 1966; Konrad Lorenz, *On Aggression,* New York, Harcourt, Brace and World, 1966; Edwin P Hollander, *Principles and Methods of Social Psychology,* New York, Oxford University Press, 1967, pp. 388-394.

6b Herman Foertsch, *The Art of Modern Warfare,* New York, Oscar Piest, 1940, p. 3.

[7] Admiral Sir Reginaid Bacon & Francis E. Mc Murtrie, *Modern Naval Strategy,* New York, Chemical Publishing Company, 1941, p. 15-16

[8] Huntington, *op.cit.,* p. 63

SEVEN

THE MILITARY THEORY OF RISKS: ALARMISM

Technical evolution causes deep-reaching changes in the military sector. But as Huntington points out, military ideology is influenced only to a small extent by this evolution. Developments in weapons technology, Huntington claims, do not alter the character of military ethic "any more than the discovery of penicillin altered medical ethics".

Military men have professional responsibility for the security of the state. This responsibility leads the military

(1) to view the state as the basic unit of political organization;
(2) to stress the continuing nature of the threats to the military security of the state and the continuing likelihood of war;
(3) to emphasize the magnitude and immediacy of the security threats;
(4) to favor the maintenance of strong, diverse, and ready military forces.[1]

As pointed out above, nationalism functions as a motive for a concern with war and war preparations. One element of nationalism, that the nation is sacred, provides a rationale for a national defense system — it establishes, in other words, the object which is threatened or may become threatened. Without such an object, there would be no reason for defense.

There is a multitude of historical examples of military alarmism. For instance, in 1948 official U.S. Army policy was that the Marshall Plan and Universal Military Training (UMT) both should be instigated as instruments in defeating Communism. Military leaders were prepared to go before the Senate Armed Services Committee to discuss the need for UMT to back up the Marshall plan. In that connection, the Army handed President Truman an intelligence report which "pictured the Soviet Army on the move", presumably to attack allied forces in Europe. President Truman went before a joint emergency session of Congress on March 17, 1948, and asked for immediate passage of the Marshall Plan and UMT (later to be defeated). Shortly after the President's address to Congress, the CIA evaluated the Army's report as false: the Soviet moves were evidently redistributing of troops to their Spring stations. Still, some military kept the possibility of a war with the Soviet Union open:

[1] Samuel P. Huntington, *The Soldier and the State*, Cambridge, Mass, 1957: Harvard University Press, pp. 64-65

in April, General Omar Bradley stated that "we are not sure" that "there is no war right away"; again in June he warned that war with Russia was a "plausible possibility".[1b]

And Vagts, writing about the pre-1914 situation in Europe, points out how the military alternative was almost the sole answer to the question how the nations could achieve national security: "armies undertook to 'sell' themselves as institutions of security to the civilians of their countries, who did not want to live dangerously, except now and then... This argument was well received by parliaments, for it served to cover up the fact that armies often created insecurity at the time they were presumed to be providing security." A number of war scares occurred in England in 1847, 1852, 1884, 1889 och 1892–93, all motivated by the "alarmists" that the country was threatened by French naval increases or invasion plans. "In each case it was subsequently disclosed that no such plans had existed." To be sure, the military were not the only alarmists: for instance, in 1847, while some observers attributed the war scare to "half-pay officers in the London clubs", others saw part of the reasons in the persuasion of "Mr, Pigou, the gunpowder maker".[1c]

[1b] John M. Swomley, Jr: *The military Establishment,* Boston 1964: Beacon Press, pp. 64 ff

[1c] Alfred Vagts, *A History of Militarism,* pp. 362–364.

It is also easy to find statements made by the Swedish military which exemplify Huntington's points. The statements occur in newspaper interviews with military leaders, in service manuals, and in reports of military investigations like the *ÖB 65* (which is a study and prognosis made by the Swedish Commander-in-Chief on the future development of the national defense). In a series of interviews in the daily newspaper *Stockholms-Tidningen* in the spring of 1965 some members of the Army High Command commented upon international politics. The Commander-in-Chief said the danger of war was the same as in 1962 (when the former defense study was conducted) and that the need for a strong defense had not changed. The Chief of Staff did not see any evident changes in the dangers of armed conflict, and the Assistant Chief of Staff thought that world tensions would last. In a more recent statement the Swedish Commander-in-Chief said that "there are points of light if you look around the world, but one has to be clear over that lights can be quickly put out. We ought to continue planning the defense according the possible cases of attack, and not according to the possibility that they will actually be carrid out. Look at Czechoslovakia!"[1d] An overview in the *ÖB 65* of the military political situation up to May 1, 1965, ends with some conclusions:

[1d] *Dagens Nyheter,* september 20, 1970.

The trends in military politics thus are characterized by
that world tension remains on a varying but high level;
that dangers of various wars will continue to exist;
that wars and causes of conlict – even in remote parts of the world where the interests of the Great Powers are affected – may affect us.[2]

These are evident examples of the general risk theory which Huntington discusses. As it is expressed in the formulations above the theory has, however, certain shortcomings which renders it useless as a

[2] *ÖB 65,* Stockholm 1965: Försvarets Bok- och Blankettförråd, p. 13.

prognosis about the future. It cannot be used for predictions concerning
(a) *whether* Sweden will become involved in a war,
(b) *with whom* Sweden will in that case be in war,
(c) *what type of war* it will be (conventional or nuclear), or
(d) approximately *when* it will start. (The report clearly has the ambition to make predictions. It speaks of "developmental tendencies" which show that "political antagonisms and suspicions between the Great Power blocks continue to exist"; on the subject of disarmament it says that there are not to be expected "any agreements of essential importance in the reasonably near future". The report also maintains that "even in the future we will have to count with a high level of tension which involves considerable dangers of war"). What is said in the theory is so elastic that at any point in time it may be used *ex post facto* to explain
(a) why a war has started (world tension is said to be on a high level; war is, therefore, a natural consequence), or
(b) why a war has *not* started (world tension is said to be varying, and it has not at the moment the temperature which is required to cause the outbreak of armed action).

On the other hand, if the theory specified *when* a possible war will start or *how long* a period of peace may be expected, it would be testable. Such clarifications are avoided in the *ÖB 65;* thus, the theory is safe for any attack on the military ability to prophesy about the future, but at the same time is useful as a device for alarmism.

Military alarmism is functional to the interests of the profession. The military risk theory, clothed in cold, analytical – but, as we have seen, not very precise – terms becomes a rationalization for the motive to maximize one's own resources. (This, of course, holds true for other professions as well.[3] It is, however, clear that since the risk theory is applied to international politics, it is a body of thought which may have consequences not associated with other professional ideologies).

One passage from Huntington summarizing the basis of the military view of threats against national security, provides a hypothesis about the relation between professionalization and the sense of danger.

> The goals of professional competence requires the military man to estimate the threat as accurately as possible. But the military man also has a professional interest and a professional duty to stress the dangers to military security. Consequently the objective realities of international politics only partially determine the military estimate of the situation. The military man's views also reflect a subjective professional bias, the strength of which depends upon his general level of professionalism. This professional bias, or sense of professional responsibility, leads him to feel that if he errs in his estimate, it should be on the side of overstating the threat. Consequently, at times he will see threats to the security of the state where actually no threats exist.[4]

The higher their level of professionalization, the more likely officers are to emphasize that there is a great risk of war. Since rank may be taken as a rough index of professionalization, we would expect officers of high rank to judge the international situation "pessimistically" more often than lower rank officers.

[3] Compare Lord Salisbury's remark: "If you believe the doctors, nothing is wholesome; if you believe the theologians, nothing is innocent; if you believe the soldiers, nothing is safe." Vagts, *A History of Militarism,* New York, 1959: Hollis & Carter, p 362

[4] Huntington, *op.cit.,* p. 66.

1. War risks: "pessimism" and "optimism"

In the survey which I conducted in the autumn of 1962 among military personnel on active duty in the Swedish armed forces the following question was included: According to your opinion, how great is the risk that there will be a great war in Europe within the next 5 years?

(Response alternatives: very small, rather small, rather great, very great, don't know.)[5]

Before discussing the results, it is worth noting that the questionnaires were sent out in the beginning of November, 1962, shortly after the culmination of the Cuban missile crisis. For this reason, the percentages below should not be taken as representative of the current opinion among the Swedish military. However, what is primarily interesting in this context is the variation in the percentages by rank. It may be assumed that these relations still hold.

As can be seen from table 7, the hypothesis of a more widespread alarmism in higher ranks is confirmed. The table lists figures for Army officers only, but the tendencies in the Air Force and the Navy (here unreported) are the same, though slightly less pronounced.

[5] Bengt Abrahamsson, *Anpassning och avgångsbenägenhet bland militärt befäl*, Stockholm 1965: Militärpsykologiska Institutet, MPI rapport nr 37.

TABLE 7
Opinions on risks of war[6]
(By rank and Army branch)

	Considers the risk of war in Europe within 5 years to be very or rather great. Per cent.			
Rank	Infantry		Other Army branches	
2. Lieutenant	37	(16)	–	(7)
1. Lieutenant	44	(64)	36	(88)
Captain	54	(80)	51	(139)
Major	60	(20)	53	(37)
Lieutenant-colonel	–	(5)	58	(14)

[6] The survey was conducted among all military personnel on active duty within the Swedish defense, except full colonels and higher ranks, which were inaccessible. A nonproportionate stratified sample was used. The data in the table are previously unpublished results from the survey. Estimates computed by Erikson's inversion test (an extension of Kendall's tau; see R. Erikson, *Yrkesval och officerrekrytering*, MPI report no 31, June 1964, appendix 8) for both branch categories give statistically significant results ($p < .05$).

Officers in the infantry consistently judge the situation more serious than officers of other Army branches; however, the differences are small. In both categories, the judgment is more alarmist in higher than in lower ranks.

As Huntington suggests, it seems adequate to view the military risk theory as an effect of the professional culture (either directly through indoctrination or, as was pointed out earlier, indirectly through the impact of selection process.) According to this interpretation wars and situations of armed conflict become more important *as theoretical problems* the more one concerns oneself with military matters. This may be assumed to lead to a propensity for analyses in terms of threat and retaliation, which in turn prepare the ground for the risk theorems in the

ÖB 65 and other documents. (In other words, the professionalization of the military man can be said to create a trained incapacity for considerations of what consequences peace and reduction of international tension may have for the national defense program.)

2. Adaptation to job and estimates on the likelihood of war

In the discussion on the military mind (above, ch. 4) it was proposed that unsuccessful socialization to the military profession is linked with feelings of maladjustment to the job, and that people leaving the military profession are characterized by attitudes that do not fit in with the organization's value system. This implies, on a general level, that badly adapted individuals also exhibit uncongenial attitudes and opinions. More specifically, if absolutism and thereby "pessimistic" war-risk estimates are attitudes perceived as functional by the military establishment, we would expect to find "pessimism" related to the individual's feelings of being well adjusted to his job situation. If subjective adjustment to the job is operationally defined as positive answers to the question "How satisfied are you, on the whole, with your military job?", and a "pessimistic" outlook is measured by answers to the question about the risks of a major war in Europe within 5 years, a positive although not very strong correlation between the two variables is obtained; the better the adjustment to the job, the higher the proportion of "pessimists" (table 8).

TABLE 8

"Optimistic" and "pessimistic" judgments on the risks of war (by level of adjustment to job). Per cent of total officer corps.

Adjustment level	Optimists	Pessimists	Total	n
1 (high)	50.0	50.0	100.0	59
2	59.6	40.4	100.0	334
3 (low)	71.2	28.8	100.0	286
Total	56.5	43.5	100.0	679

Computation of gamma yields a positive correlation of .23 ($p < .05$).

If rank is held constant, there is no tendency for this relationship to disappear. Thus, for each rank it seems that successful adaptation to the military profession involves the internalization of a "pessimistic" outlook on the security of the state.

The process of professional socialization may be interpreted as the gradual learning of conforming to the values of the profession. Probably, the covariation found between "pessimism" and subjective adjustment is an indicator of the professional members' responses towards professional social control in the alarmist direction.

3. Age and "pessimistic" outlook: some civilian data

It has been argued that the relationship between rank and alarmism mirrors a generally greater proneness of older people to stress international unrest and threats towards the nation. Should this be the case, there is obviously ground for serious doubts about the existence of a special alarmist element in the military professional culture.

Data on attitudes among the normal population, exactly comparable to those in the 1962 Swedish military survey are lacking. However, the Swedish Bureau of Psychological Defense (Beredskapsnämnden för psykologiskt försvar) in 1965 conducted a survey on a national sample, which included the following question:

How great do you consider the danger being that the present unrest in the world will become widened into a conflict between the Great Powers, a conflict which will also involve Europe?[7]

Details on the total percentage distributions are left out here, since the interesting issue is the relation of age to "pessimism". The differences between age classes are small, but there is a slight trend towards greater proportions of "pessimists" among the *younger* people in the sample. The percentage of people saying that the risks of a major war are "very great" or "rather great" among age classes 15–19 is 54; among people aged 20–24, 51 %; in the older age categories the percentages vary between 41 and 44.

There is nothing in these figures to support the notion of a more wide-spread alarmist outlook on world politics among older people; needless to say, this gives further support to the hypothesis that adaptation to the military profession is most successfully achieved if the profession's basic values as to the risks of war are accepted.

[7] Kurt Törnquist, *Försvarsvilja och bedömning av krigsrisker* (Defense Motivation and Judgments on the Risks of War) Stockholm 1966; Beredskapsnämnden för psykologiskt försvar, report no. 28, mimeo.

EIGHT

AUTHORITARIANISM

The writings on the military mind fairly often reflect the idea that since the military organization is authoritarian rather than democratic, it fosters an authoritarian outlook. In an explanation of the relationship between leadership style and the prerequisites of military organization, Commander H. E. Smith notes:

> The military mind is characterized as impatient with dissent, inexpert in the art of persuasion, and ignorant of the importance of minorities in our land. The democratic given-and-take, it is said, is missing in military life, the frictions of debate are not tolerated. The military mind is therefore not pliant. It is given to regimentation of people and ideas, being intolerant of disagreement.
>
> Military life is undoubtedly more ordered than civilian affairs. To me, it is a refreshing attribute of military men that they follow orders. The business of preparedness is so complex and costly that it had better be ordered. When there is a clearly defined chain of command and an individual who has the authority and responsibility for making a decision, military men will argue before him with vigor, in spite of what their critics say, but when the decision is made, they will loyally abide by it.[1]

In a more critical manner, Hanson Baldwin has argued that the military mind is "rational" and "pragmatic", but not intuitive, being well able to grapple with tangibles but less well with intangibles. "The military mind...often is not fundamentally a democratic mind. No military organization can be democratic and still retain its essential military characteristics; the two are antithetical."[2] Likewise, Justice William O. Douglas describes it as not knowing "the give and take of public debate, the art of persuasion of people, the value and importance of dissent and disagreement".[3]

In the sociological literature, the military establishment has often been described in terms similar to those quoted.[4] It is possible, however, that with the increase in the proportion of managerial positions, the military culture is successively being transformed in a more non-authoritarian direction, where less premium is put on absolute obedience and the leadership forms are manipulative and persuasive rather than dominative; this is a wellknown and central hypothesis in Janowitz's analysis of the American military.[5]

The extent to which authoritarianism is functional for the carrying out

[1] "What Is the Military Mind?" *U.S. Naval Institute Proceedings*, vol. 79, May, 1953, p. 511

[2] Hanson W. Baldwin, "When the Big Guns Speak", in Lester Markel (ed.), *Public Opinion and Foreign Policy*, New York, Harper & Bros., 1949, p. 119.

[3] "Should We Fear the Military?", *Look*, vol. 16, March 11, 1952

[4] Cf. e.g. Stouffer, S., et al., *The American Soldier*, I, Princeton University Press, 1949, p. 55; Husén, Torsten, *Militärt och civilt*, Stockholm, Norstedts, 1956, pp. 43-44; Gunnar Boalt, "Militärt och civilt. Några sociologiska synpunkter", *Social årsbok, 1950-51*, Stockholm 1951; KF, pp. 9-27

[5] *The Professional Soldier*, Glencoe, Illinois, 1960: Free Press pp. 8-9

of daily tasks within the military establishment most probably varies between occupational specialties and between ranks, with relatively low-ranking operational positions presumably being more conducive to authoritarian behavior than, for instance, higher ranking technical and supply positions. To the extent that the proportion of these tasks differ between services we would also expect measurements of authoritarianism to show divergencies. This may explain why there seems to be no relationship between F-scores and attitudes towards choosing the Air Force as a lifetime career.[6] In the Air Force, the tasks typically involve flying and/or support functions, and the daily job has relatively little to do with the command of large units of men in drill and combat training.

In the Army, however, the situation is different, and it would not be unreasonable to expect that persons choosing that service as a career would tend to be relatively authoritarian. A study by Korpi on the Swedish military[7] by and large confirms the notion that recruitment to the Army tends to overrepresent persons with authoritarian outlooks. Investigating army NCO aspirants with respect to degree of authoritarianism, he found relatively consistent trends in the direction of higher authoritarianism among those who *planned* to attend cadet school (in comparison with those who did not), those who actually *attended* cadet school (as against those who did not), those who chose officer as their *primary career choice* (in comparison with those who chose officer as an secondary alternative or not at all), and those who actually *began* a military career (compared with those who did not). The sole exception to these fairly coherent findings was the fact that NCO candidates who already at induction chose officership as a career were not significantly different from the rest of the NCO candidates with respect to authoritarianism. This may, however, be explained by the fact that these early-choosing inductees possibly differ from the others in terms of, for instance, intelligence score. If the early-choosing inductees constitute a subsample with higher-than-average intelligence, it should follow that they would have a lower-than-average degree of authoritarianism.[8]

It should also be noted that Korpi deliberately oversampled NCO's from Army branches where the authoritarian aspects of the military system may be assumed to appear in a more "pure" form than in others. Men from the infantry constitute about 72 per cent of the sample, as against 63 per cent of the total population of NCO trainees.[9]

In conclusion, differences in authoritarianism between branches seems at least partly to be a product of the less technical and more "heroic" type of tasks which await the prospective officer in the Army compared to, for instance, the Air Force.

Authoritarianism, whenever it is found among military officers, is not only a product of long-term professional socialization and indoctrination. As we have seen from Korpi's data, authoritarianism is related to positive attitudes to officership as a profession even *before* beginning actual officer training. In terms of Rosenberg's model for occupational choice, the professional values are in this case the products of anticipatory

---

[6] French, E.G., and Ernest, R.B., "The relationship between authoritarianism and acceptance of military ideology", *Journal of Personality*, 1955, vol. 24, pp. 181-191. Cf. also Campbell & McCormark, who found a significant *decrease* in F-scores among airacadets after one year of training. (Campbell, D.T., and McCormack. T.H., "Military experience and attitudes toward authority", *American Journal of Sociology*, 1957, 62, 482-490

[7] W. Korpi, *Social Pressures and Attitudes in Military Training*, Stockholm 1964: Almqvist & Wiksell, esp. ch 6.

[8] Korpi notes that those selected at induction have, more often than others, higher secondary school education *(Ibid.,* p. 86). This supports the notion that the early choosers are a relatively biased subgroup of the whole group of NCO's to be, in terms of intellectual criteria.

[9] *Ibid.,* p. 40

socialization, or of the fact that one chooses an occupation that he likes and where he feels that his attitudes will "fit".

However, to the extent that the military organization becomes highly technical and its techniques of command shift toward manipulation rather than domination, the emphasis on authoritarianism can be expected to gradually diminish. Since it is derived more from the specific forms of *organization* rather than from the imperatives of professional *goals,* it is more subject to change than the rest of the factors discussed above; for instance, while alterations in the traditional dominative leadership principles probably will affect the "market" for authoritarian attitudes and behavior, such changes will have no or very minor effects on political outlooks or the estimates of the probability of war. The latter factors are essentially independent of organizational *forms,* but heavily dependent of organizational *goals.*

NINE

SOME IMPLICATIONS FOR MILITARY OCCUPATIONAL PRESTIGE AND INTELLECTUAL RECRUITMENT

On the general assumption that potential candidates to an occupation tend to match their impressions of it with their own values, we expect the recruitment to an occupation to be related to its image among the general population, and particularly among young people. It is a commonplace that in a certain society at a certain time some values are less prevalent that others, and some outlooks less popular than others.

In the modern, industrialized and largely liberal Western society a value such as nationalism — the foremost ideological base of the military profession — tends to fluctuate with the patterns of international conflict: the stronger the threats — real or perceived — to the country, the greater the spread and intensity of nationalism. The military establishment typically finds recruitment less difficult during periods of emergency; for example, officer recruitment in Sweden during the first half of the 1940's was generally satisfactory. It deteriorated both qualitatively and quantitatively in the period immediately following the war.[1]

Thus during such periods it appears that national sentiments to a large extent counterbalance the otherwise unfavorable images of the military among civilians. There can be no doubt that such negative images do exist, sometimes highly pronounced. As Janowitz has noted with regard to the United States, officership remains a relatively low-status profession, in contrast to the public acclaim accorded invidual military heroes.

> The results of a national sampling of opinion in 1955 place the prestige of an officer in the armed services not only below that of physician, scientist, college professor, and minister, but also below that of the public school teacher.[2]

On the sheer basis of the close correspondence of prestige measurements taken in different countries[3] we would expect similar results in other industrialized nations. The findings have been replicated, for instance, in Sweden, where the position of colonel in the armed forces was rated lower than a teacher in elementary or secondary school, head cashier, and pharmacist. Non-commissioned officers ranked roughly equivalent to shop salesman, but lower than policeman and taxicab owner.[4] The respondents in the study cited were asked to indicate what

---

[1] *Befälsordningen vid infanteriet (SOU 1953:28)*, Stockholm 1953, pp. 30-36

[2] Janowitz, *The Professional Soldier*, Glencoe, Illinois, 1960: Free Press, pp. 3-4. Janowitz is referring to the study *Attitudes of Adult Civilians Toward the Military Services as a Career*, Washington, 1955: Public Opinion Surveys, Inc.

[3] Cf. e.g. Alex Inkeles and Peter H. Rossi, "National Comparisons of Occupational Prestige", *American Journal of Sociology*, vol. 61, 1955-56, pp. 329-339.

[4] Gösta Carlsson, *Social Mobility and Class Structure*, Lund, 1958: Gleerups, pp. 147 ff.

[5] *Ibid.*, p. 149

parents in general would think of the indicated occupation as an occupation for a son of theirs. According to the author of that study, this might explain the low ranking of the two military occupations. "War and military preparations are no more pleasant to contemplate in Sweden than in any other country, and people did therefore not think of the occupations as attractive ones, or did not think that other people thought them attractive".[5] However, even if the questionnaire item had been rephrased to measure the image of prestige more directly, one may doubt that the two military ratings would have changed very much. As another study has found, even when military men themselves rank a number of occupations, sergeant is rated clearly lower than grade school teacher, and captain in the army lower than high-school teacher.[6]

Strongly negative images of the military occupations also prevail among the population of potential recruitees to the military profession in Sweden. In a survey among senior high school students in Stockholm in 1967, the respondents were asked to classify a number of occupations on differential scales, based on adjectives like helpful–unhelpful, nice–not nice, modern–unmodern, valuable–valueless, powerful–powerless, high prestige–low prestige, impopular–popular, and so forth. The students placed the military occupational categories of army captain and colonel at the negative end of most scales; they were, in comparison with all other occupations named in the study,[7] judged as less helpful, less nice, more valueless, having less prestige, being more passive, unfriendly useless, unsatisfied, ignorant, impopular, and lazy. Only minister was named as more unmodern than captain and colonel; the colonel was classified as the most powerful, while captain in this respect ranked between high-school teacher and dentist; on intelligence, captain was classified as the least intelligent, followed by minister and colonel, with physician as the most intelligent. The military titles were placed in the middle of the "practical–unpractical" continuum and judged to be less practical than the physician and the dentist, but more practical than the lawyer, high-school teacher, and minister.[8]

In view of such perceptions of the military profession among pontential candidates of officer positions, what inferences can be drawn regarding trends in intellectual recruitment? First, it should be noted that the traditional conception of the military occupations as involving demands on physical strength and high potentiality for action, rather than scholastic achievement, would tend to make the military relatively attractive in the eyes of men with comparatively poor educational performance (and conversely, relatively non-attractive to the others). Second, negative impressions such as the ones referred to above will tend to lower the motivation to seek military employment: hence the military profession, in comparison to others more positively perceived, will have to lower its intellectual requirements in order to get the necessary manpower.[9] In a situation of full (or next to full) civilian employment – so that the economic motive of entering the military is reduced and occupational choice is relatively free – we expect those who choose a

[6] Bengt Abrahamsson, *Anpassning och avgångsbenägenhet bland militärt befäl* Stockholm, 1965: Militärpsykologiska Institutet, rapport no. 37. The position of colonel was not rated in this study.

[7] These were minister, high-school teacher, lawyer, physician, and dentist.

[8] Olof Fràndén, *Officersyrkets anseende bland gymnasister i Stockholm,* Stockholm, 1968: Militärpsykologiska institutet, B-report no. 8 (mimeo).

[9] The official report *Befälsordningen vid infanteriet (SOU 1953:28)* mentions that the requirements in mathematics appear to have been one source of the difficulty in recruiting officers for the technical branches of the Swedish army; however, in 1952, these requirements were lowered after which the reporting committee estimates that "this ought to lead to an improvement of the recruitment to these branches, at least quantitatively" *(Ibid., p. 32)*

military occupation to have lower intelligence ratings (and lower school grades) than those who select other careers.

A number of investigations in Sweden from the mid-1940's and onwards support this assumption. From the results of a study of grades to the Higher Certificate Examination (studentexamen) carried out immediately after World War II Colonel Lemmel concluded that "the students who have chosen the occupation of officer have an average grade in the HCE which not immaterially falls below the average mark of those students of the same age who have chosen other occupations."[10] In the early 1960's a survey by Erikson showed that "those who want employment as officers have, on the average, lower intelligence than those... who do not/want to become officers/".[11]

Fråndén in a recent paper shows that the average marks on the HCE have been consistently lower for students at military colleges than for other college students during the period of 1956–62, and that the trend has been similar ever since the beginning of the 1930's.[12] These findings do not seem to be limited to Sweden, since some French data also demonstrate a similar tendency.[13]

The consistency in these findings seem to point at the fact that the military profession – in marked contrast to the professions of law, medicine, theology, and several others[14] has had very little success in attracting the intellectually superior among the young generation, and has had marked difficulties in recruiting even the moderately bright.

Professionalization and the military mind: Summary and conclusions

Part II has presented (a) a definition and a rationale for the use of the concept of military mind (chs. 3 and 4); (b) some hypotheses regarding its major components (chs. 5–8); and, (c) a discussion of the implications for military prestige and intellectual recruitment (ch. 9). Some data, although scarce, have been presented in support of the arguments.

1. The concept of military mind appears fruitful for a number of reasons.
1.1 There is substantial intersubjective agreement among writers of different persuasions that relatively stable clusters of specific attitudes exist among members of the military profession. If so, and to the extent that these attitudes are valid indicators of behavior, predictions regarding the behavior of the military *as a group* may be made.
1.2 The military mind may be regarded as the joint product of various homogenization processes, such as anticipatory socialization, selection at induction, selection in promotion, and professional socialization. For given individuals, we expect these processes gradually to produce increasing crystallization of attitudes, thus increasing the accuracy of predictions the more professionalized the individuals. Consequently, in the military field, rank may be taken as an index of professionalization$_2$.

[10] "Rekryteringen av arméns aktiva officerskår ur intellektuellt kvalitativ synpunkt", by C.F. Lemmel. Appendix to *Befälsordningen vid infanteriet (SOU 1953:28)*, pp. 203-217. Quotation on pp. 215-216.

[11] Robert Erikson, *Yrkesval och officersrekrytering*, Stockholm 1964: Militärpsykologiska institutet, report no. 31

[12] Olof Fråndén, "Notes on mobility into and out of the Swedish officer corps", in Jacques van Doorn, ed., *Military Profession and Military Regimes*, The Hague, 1969: Motton, pp. 107-126.

[13] "Attitudes et motivation des candidats aux grandes écoles", *Revue Francaise de Sociologie*, vol. 2, pp. 133 ff.

[14] Cf. Fråndén "Notes...", p. 123, table 15.

1.3 The *content* of the homogenization processes will obviously differ between professions. As a general assumption, we may expect the various professional minds to be related to the particular goals and values of the respective professional groups. Investigating such goals and values, then, may yield insights into the "internal life" of those various groups, and form a basis for the comparative analysis of professions.

2. The major components of the military mind were identified as (1) nationalism, (2) pessimistic beliefs on human nature, (3) alarmism (i.e., pessimistic estimates on the probability of war), (4) political conservatism, and (5) authoritarianism. Components (1), (2), and (3) were assumed to derive from the goals and values of the profession, and may hence be expected to be independent of the internal structural changes of the military organization; whereas (5) can be assumed to be dependent on the specific forms of organization, and therefore expected to vary according to the type of military activity (being less in technical and "non-heroic" types of job). (Political attitudes will be left for later consideration (ch. 10)).

Empirical data from the Bundeswehr showed nationalism to be more prevalent among higher than among lower ranks and also higher among military personnel than among the civilian population. Swedish data showed pessimism about the possibilities that man can learn to avoid war to be more common among reserve officers than among civilian students, although the differences were small. Swedish data also showed alarmism to be more wide-spread among higher than lower ranks. No relationship between age and alarmism was found among the population at large, supporting the notion that alarmist attitudes are a function of processes of professional homogenization. Further, Swedish young men seeking military employment were found to exhibit authoritarian attitudes more often than comparative groups of men not doing so.

Although political attitudes will be discussed in detail later, it should be mentioned here that the sample of Swedish officers shows that *both* alarmism and political conservatism tend to be more common the higher the rank. This supports the assumption of increasing "crystallization" of attitudes with increasing degree of professionalization.

3. Finally, a series of surveys in Sweden have shown officer candidates to be inferior in intellectual quality to students at civilian colleges. (Comparable findings are reported from France). This may be explained by:

3.1 The low degree of external threats against Sweden during the post-war period, keeping nationalist sentiments at a low level and making it less likely that military professional skills will actually be used; i.e., military employment has an "idleness prognosis" which would appear to make it less attractive.

3.2 The situation of full employment, in combination with relatively low pay-rates for military jobs.

3.3 The improvement of general education, opening educational opportunities to young men who previously would have relied on military schools for free secondary and higher education.
3.4 The low prestige of the military (resulting from 3.1–3.3), with unfavorable stereotypes creating negative intellectual recruitment, and this in turn creating unfavorable stereotypes, etc, in feedback cycles.

Postscript

Shortly before the publication of the present work, an article appeared which lends some support to item 1.2 above. H. Tromp of the Polemological Insitute of the University of Groningen administrated Rokeach's Dogmatism scale to two samples of respondents, one military and one civilian (indicated by Tromp to have been comparable on a number of basic background variables). The author reports, first, that the two groups differed clearly with regard to D-scores, the military group scoring higher; second, that the military personnel scored higher on six selected items measuring "militarism" and "hawkishness"; and third, that cadets from the Royal Military Academy of the Netherlands on these items scored higher than a group of conscripts, and the conscripts higher than a group of "pure" civilian respondents, the results thus strengthening the hypothesis of a positive correlation between level of military professionalization and D-score (the latter reported in another study by Eckhardt to correlate positively with variables such as nationalism, anti-Communism, conservatism, and authoritarianism). (See H. Tromp, *The Assessment of the Military Mind*, paper presented at the Social Science Symposium, Ile de Bendor, 20–25 Sept., 1970).

PART III

PROFESSIONALIZATION, INFLUENCE AND POWER

TEN

ELEMENTS OF MILITARY CONSERVATISM: TRADITIONAL AND MODERN

Introduction

One of the most commonly noted characteristics of the military mind is a marked conservatism. It forms a cornerstone in Huntington's theory of civil-military relations;[1] it is a recurring theme in Alfred Vagts's historical account of militaristic thinking, and in Gordon Craig's analysis of the political interventionism of the Prussian military;[2] and it is a main element in Janowitz's treatment of the identity and ideology of American military leaders.[3]

Opinion surveys of the American, West German, and Swedish officer corps lend strong support to the hypothesis of military conservatism (table 9). The data show that as far as the Bundeswehr officers and their Swedish colleagues are concerned, they deviate strongly in their political opinions from those of the population at large. No comparable population figures are available for the United States but, as Janowitz points out, "it is clear that higher education is associated with a greater concentration of liberal attitudes in the population at large. Thus, the emphasis on conservative attachments is even more noteworthy, since (US military officers) constitute a group in which higher education does not weaken conservative orientations."[4]

It may be argued that officers show conservative attitudes because they constitute a socially privileged group; as is well known, professionals are more often conservative than liberal in their political tendencies.[5] This can be only a partial explanation, however, since occupational groups with an even more pronounced elitist character often show less conservative preferences. For instance, data on voting patterns among big businessmen, business managers, high officials in private service and higher grade civil servants in the 1964 election to the Swedish second chamber show that only 46 per cent voted conservative.[6] This may be compared to the military data collected in the 1962 survey. The fact that 85 per cent of the officers preferred the Conservative party suggests that the military group is strikingly homogeneous in its political sympathies. The high proportion of conservative officers in the Swedish military is even more noteworthy given its relatively egalitarian recruitment (59 per cent middle or lower class).[7] A pattern similar to the Swedish one is exhibited by data from West Germany, showing a much higher

[1] Samuel P. Huntington, *The Soldier and the State,* Cambridge, Mass., 1957: Harvard University Press

[2] Alfred Vagts, *A History of Militarism,* New York, 1959: Meridian Books, esp. chs. 9 and 12. Gordon A. Craig, *The Politics of the Prussian Army,* Oxford, 1955: Clarendon Press

[3] Morris Janowitz, *The Professional Soldier,* Glencoe, Ill., 1960: Free Press, ch. 12

[4] *Ibid.,* p. 238

[5] See, for instance, Seymour M. Lipset and Mildred A. Schwarz, "The Politics of Professionals", in H.M. Vollmer and D.M. Mills, *Professionalization,* Englewodd Cliffs, N.J.: Prentice-Hall, pp. 299–310

[6] *Riksdagsmannavalen åren 1961–1964: II,* Stockholm, 1965: Statistiska Centralbyrån, p. 95

[7] Bengt Abrahamsson, *Anpassning och avgångsbenägenhet bland militärt befäl,* Stockholm, 1965: Militärpsykologiska institutet. MPI rapport no. 37, p. 61

TABLE 9
Military political preferences

Political identifications among U.S. Army officers. Year: 1954.[a] Per cent

	Conservative	Somewhat conservative	Somewhat liberal	Liberal	No answer	Total	Number
Academy Graduate	27.4	42.1	18.9	6.3	5.3	100.0	95
Non-Academy Graduate	23.3	46.6	23.3	4.3	2.5	100.0	116

Political identifications among soldiers in the Bundeswehr Year: 1960.[b] Per cent
Question: Which of the political parties comes closest to your own position?

	CDU/CSU	FDP	SPD	Other parties	No information	Total	Number
Recruit	47.5	6.4	24.0	3.7	18.0	99.6	1 302
Corporal	55.7	3.6	24.1	1.9	14.4	99.7	1 509
Sergeant	74.2	3.2	13.1	3.5	5.6	99.6	595
Senior NCO	76.9	4.3	11.6	3.3	3.5	99.6	309
Officer candidate	73.5	11.4	7.3	2.7	4.7	99.6	612
Officer (Lt Capt)	74.6	11.5	4.8	3.9	4.8	99.6	108
General population, 1964	34	6	34	2	24	100.0	
Self-employed and professionals, 1964	40	12	19	2	27	100.0	

Political identifications among Swedish officers. Year: 1962.[c] Per cent
Question: With which political party do you sympathize the most?

	Conservatives	Center Party (Agrarians)	Liberals	Social Democrats	Communists	Total	Number
Officers	85.1	1.7	9.3	3.9	0.0	100.0	781
Electorate, local election, 1962	15.5	13.1	17.1	50.5	3.8	100.0	

Conservative preferences among Swedish Army officers. By .ank. Year: 1962.[d] Per cent
Base numbers within parentheses. Per cent sympathizing with Conservatives.

	Second lieutenant		First lieutenant		Captain		Major	
Army: infantry	60	(16)	86	(64)	90	(80)	95	(20)
Army: other branches	–	(7)	81	(88)	86	(139)	94	(37)

a) *Source*: Janowitz, *The Professional Soldier*, p. 240 (table 29). These figures relatively accurately also represent political identifications among Pentagon officers (1954) and among Navy and Air Force officers. Cf. *ibid.*, pp. 236–241.

b) *Sources*: Military, Eric Waldman, *The Goose-Step Is Verboten*, Glencoe, Ill.. Free Press, 1964, p. 216, table 4. Population, Noelle and Neumann, *Jahrbuch der öffentlichen Meinung, 1958–1964*, pp. 3 ff, and ENID, *Informationen*, Nos. 13 and 26 (1964). Quoted from Lewis J. Edinger, *Politics in Germany*, Boston, 1968: Little, Brown, and Company.

c) *Source*: Abrahamsson, *Anpassning och avgångsbenägenhet bland militärt befäl*, p. 69

d) *Source*: Bengt Abrahamsson, "The Ideology of an Elite", in J.A.A. van Doorn (ed.), *Armed Forces and Society*, The Hague, 1968: Mouton, p. 75.

proportion of CDU/CSU sympathizers among Bundeswehr officers than among civilian self-employed and professionals (see table 9).

Explanations of why conservatism is particularly attractive to the military can be based only to a limited extent on social position; very few inferences with regard to political attitudes can be drawn from data on social background. Instead, we have to ask: how is it possible that such a high proportion of the officers have conservative sympathies when we know that the military is one of the most broadly recruited of professions? To answer this, we have to take into account, first, the set of values that are functional to military professional performance, second, the historical traditions of the military and, third, the relation of the military to other influential groups, particularly business and industrial elites.

1. Functional attitudes

As has been noted by Herbert McClosky, among others, the notion of man as fundamentally aggressive is an inherent part of conservative ideology. McClosky, in an article on the elements of conservatism, with reference to the works of Edmund Burke, Russell Kirk, Clinton Rossiter, and others,[8] points to the agreement among them on the treatment of a number of topics, some of which are:

> Man is a creature of appetite and will, "governed more by emotion than by reason" (Kirk), in whom "wickedness, unreason, and the urge to violence lurk always behind the curtain of civilized behavior" (Rossiter). He is a fallen creature, doomed to imperfection, and inclined to license and anarchy.[9]

Since society cannot be held together under such generally hostile conditions, it has to impose certain restrictions on the individuals:

> Order, authority, and community are the primary defense against the impulse of violence and anarchy. The superiority of duties over rights and the need to strengthen the stabilizing institutions of society, especially the church, the family, and, above all, private property.

McClosky supports his analysis with data from a sample of Minnesota residents. Conservatism in shown to correlate with certain personality variables, like dominance, alienation, hostility, and pessimism. Conservatives, oftener than liberals, tended to endorse items like "duties are more important than rights", "you can't change human nature", "the world is too complicated to be understood by anyone but experts", and so forth. These tendencies remained when such factors as education, occupation, socio-economic status, and possible response set were controlled for. McClosky summarizes:

> These tendencies may also lie at the root of the conservative inclination to regulate and control man; to ensure that he will not violate the conditions necessary for order; to train him to value duty, obedience, and conformity; and to surround him with stabilizing influences, like property, church, and the family. The high values placed on authority, leadership, and natural hierarchy, and on an elite to guide and check the rest of mankind, apparently derive from the same set of psychological impulses.

[8] Burke, *Reflections on the Revolution in France,* New York, 1962: Holt, Rinehart and Winston; Russell Kirk, *The Conservative Mind,* Chicago, 1953; Clinton Rossiter, *Conservatism in America,* New York, 1955

[9] Herbert McClosky, "Conservatism and Personality", *American Political Science Review,* 1958, vol. 52, pp. 27–45.

McClosky's examination of the conservative syndrome coincides well with Huntington's analysis of the military mind, which is echoed also by Coates and Pellegrin in their *Military Sociology*:[10]

> /T/he military man concedes that he holds a pessimistic view of the nature of man. He believes, however, that his conception is the correct one. As Huntington explains it, the military view of man emphasizes his propensity for conflict, violence is "rooted in the permanent biological and psychological nature of men". Man is weak, selfish, irrational. . . . Man being what he is, conflict is inevitable.[11]

Low faith in people has been shown by Rosenberg to correlate with pessimistic attitudes towards the possibilities of avoiding war.[12] Table 10 shows a cross-classification of the variable "faith in people" with answers to the statement "The most we can hope to accomplish is the partial elimination of war".

[10] Charles H. Coates and Roland J. Pellegrin, *Military Sociology: A Study of American Military Institutuions and Military Life*, University Park, Maryland: The Social Science Press, 1965

[11] *Ibid.*, p. 50

[12] Morris Rosenberg, *Occupations and Values*, Glencoe, Ill., 1957: Free Press

TABLE 10

	High		*Faith in people*			Low
	1	2	3	4	5	6
Agree	32	39	43	50	52	59
Disagree	51	43	44	40	33	36
Undecided	17	18	13	10	15	5
Total	100	100	100	100	100	100

Source: Rosenberg, *op.cit.*, table 9

The lower the position on the "faith" variable, the higher the proportion of "pessimists" on the matter of elimination of war.

Thus far, the data cited have served only to demonstrate the general interrelationships of conservatism with other variables for non-military respondents. But for professionals who are preoccupied with matters of war and war preparations, a pessimistic view of man and of the prospects of avoiding war seems to be fundamentally in accordance with daily occupational practices, and McClosky's findings seem relevant to the military setting. Conversely, an optimistic view of the possibilities of bringing about disarmament would threaten the basic professional values of the military. As has been shown earlier, alarmism (i.e., the belief in the high probability of war) among officers is positively related to military rank and to work adjustment, supporting the notion that a "pessimistic" view tends to facilitate performance in the military role.[13] And the study on Swedish reserve officers showed that they less often than comparable groups agree to the statement "Man can learn to avoid war".[14]

McClosky also suggested that there is "a conservative inclination to regulate and control man; to ensure that he will not violate the conditions necessary for order; to train him to value duty, obedience, and conformity". The functional importance of such attitudes to the military role is substantiated by Korpi's study showing that Swedish

[13] See ch. 7.

[14] See above, ch. 6.

Army NCO aspirants who planned to attend cadet school or who actually attented such schools were more authoritarian than comparable groups of non-aspirants.[15]

[15] See above, ch. 8.

To summarize, military men as a professional group seem to be conservative partly because many of the values and attitudes that are part of the conservative syndrome appear to facilitate, and tend to support, adjustment to the professional role. Conversely, high faith in the rationality and reason of man, the belief that man is capable of eliminating war, and an "optimistic" view in international relations tend to be incompatible with successful professional performance.

2. Associations with ruling elites

The most typical and frequent explanation of the military's conservatism emanates from its characteristic of being associated with the ruling groups in the society. As we have seen, this is part of Mosca's analysis of civil-military relations in Europe and the United States; a more recent example is provided by Andreski.[16] As a variety of historical examples show, ruling elites and the military often agree on the necessity of the status quo, and in being suspicious and hostile against rapid social transformations. This point is well brought out by Alfred Vagts, when he asserts that the military traditionally has not been in the forefronts of revolutions, but rather has chosen to support the existing power structure.

[16] Gaetano Mosca, *The Ruling Class*, New York and London, 1939: McGraw–Hill. Translated, edited, and revised by Arthur Livingstone. Cf. also S. Andreski, "Conservatism and Radicalism of the Military", *European Journal of Sociology*, 1961,1.

As it takes many years to organize and equip an army, a long-time stability is necessary in the structure and functions of the society in which the preparation of the army is made. Hence the army by the very nature of things depends for its existence, honor, emoluments, and privileges upon the order in which it takes form; and in self-defense, if nothing more, it is conservative in relation to the order in which it thrives, whether that order be agrarian, capitalistic, or communistic.

/The great modern revolutions have been foreign to and remote from armies; armies have been closely associated with suppressions, reactions, and counterrevolutions/.

The resistance of armies and their leaders to such revolutions is based to some extent on their professional conservatism, which loves order, sees order merely in the established arrangements, and cannot discern a new one amid popular turmoil. ... Clinging to tradition, which in itself is a means for him to maintain authority, the soldier is averse to acknowledging and embracing changes forced upon him, particularly if they do not appear to favor his immediate interests.[17]

[17] Vagts, *op.cit.*, p. 30

The major internal values of the profession are in accord with the basic elements of the conservative tradition. These values stem from the feudal heritage of the military, its historical role as a guardian of the status quo, and its traditional ties with ruling establishments bent on preserving rather than changing existing social and political patterns. Many officers of the old French army chose to emigrate and to join the forces of the First Coalition, fighting the Revolution from outside; others stayed behind to work against the Revolution from within the regiments.[18] And the officers of the Tsarist Russian Army, in spite of the fact that the expansion during the First World War had brought in a large influx from

[18] *Ibid.*, p. 106

the peasant class and the intelligentia, were more often to be found among the ranks of the White than of the Red Army during the Civil War.[19]

When the fascist movements started to grow strong in Europe, officers were often eager to support them, since the policies favored by fascist leaders brought a number of seeming advantages to the military establishment.

> Fascism fulfilled many an old daydream of officers everywhere; for instance, it abolished the politician, with his awkward queries on budgetary matters, and at the same time it effectively stopped criticism on the conduct of military affairs through press or books or from the platform. ... A further lure was the fact that military men were conceded the highest rank in society under fascism; military institutions furnished the examples for all other organizations in the state. ... Besides such apparent conveniences to army men, fascism provided enlarged armies, bringing rapid promotion.[20]

Although traditionally rightist movements and ideas have had a much greater appeal to military men than the movements and ideas of the left, there are some obvious exceptions. In countries where the struggle against landholding, industrial, and/or aristocratic elites has been carried out by armed revolutionary forces, the military leaders of these forces will typically exhibit a less conservative inclination than in countries where change has been more gradual and where it does not involve any part of the armies in the overthrow of long existing regimes. Thus, the military organizations of revolutionary France and of Russia during and immediately after the revolution contained considerable progressive elements.

But again, once the revolutionary struggle is over and the period of consolidation sets in, armies tend to acquire the role of supporters of the status quo, and the officer corps tends to become part of the social establishment. As Vagts says, the army depends "upon the order in which it takes form"; and when that order becomes consolidated, the army will be its primary support.

An officer corps emerging from revolutionary war campaigns will be relatively unprofessionalized, partly because of its need to recruit officers from all possible groups, and partly because the joint revolutionary-military role counteracts the isolation of the military and its inclinations toward corporate autonomy. A case may therefore be made that professionalism and conservatism tend to develop together, and that professionalized post-revolutionary armies tend to loose their political heritage. For instance, according to Garthoff, there are a number of similarities between the Imperial Russian Army and the Soviet Army in 1935–65, both being more professionalized than the Red Army of the 1920's and early 1930's. Beginning with the mid–1930's, a number of demands from the new officer class were met. "In 1935, ranks were restored, except for general officer grades, which were not given until 1940; moreover, the creation of the rank of marshal was an "imperial" addition that had not even been found in the pre-revolutionary army. New salary scales not only favored the military, but also reflected the growing gap between officers and enlisted men, and between junior and

---

[19] "During the Civil War, between 50.000 and 100.000 officers of the old army were taken into the new Red Army (while some 200.000 entered the White Armies). ... By August, 1920, over 48.000 former officers of the Imperial Army had been taken into the Red Army — though, of course, many of these were wartime promoted officers. At the end of the Civil War, however, a large number of these officers were dismissed" (figure given for dismissals in 1921: 37 954). Raymond L. Garthoff, "The Military in Russia, 1861–1965", in van Doorn, J.A.A. (ed.), *Armed Forces and Society*, p. 246. Also compare P. Zhilin, "The Armed Forces of the Societ State", in van Doorn (ed), *Military Profession and Military Regimes*, pp. 157–174.

[20] Vagts, *op.cit.*, p. 411

senior officers. Hundreds of special stores, theaters, and clubs were established for military officers and their families. The emerging new caste of officers were given lessons in French, in polo, in dancing, and in the social graces".[21] In 1942 the *pogony* (golden epaulettes) were reintroduced, and officers were granted the opportunity to send their sons to new exclusive cadet schools, established in 1943.[22]

[21] Garthoff, *op.cit.*, p. 248

[22] *Ibid.*, p. 250

In the face of such preferences for at least a partial return to earlier practices and privileges, intensive political education is employed as a means of retaining the revolutionary heritage. In the Soviet Army today, about twenty per cent of the soldiers' training time is devoted to political instruction.[23] Almost all officers (93 per cent) are members of either the Communist Party or the Komsomol.[24] But in spite of this, as the revolutionary experience recedes into history and as new technological and highly complex demands are put on the Soviet military, it is not unlikely that it will continue "to develop as a technically-oriented, highly professional, and essentially apolitical instrument of the state";[25] "apolitical" in this context being synonymous with "de-revolutionized".

[23] *Ibid.*, p. 252

[24] Zhilin, *op.cit.*, p. 167

[25] Garthoff, *op.cit.*, p. 256

To sum up, although armies may be successful instruments for popular revolutions, once their leaders develop into a professionalized officer corps they will be less than effective as a force for social change. The officer in a professional army is prone to prefer a political status quo and to view with suspicion attempts to rapidly transform the social structure. As Huntington has pointed out with regard to the US military, professionalism fits well into a conservative political setting. "Only an environment which is sympathetically conservative will permit American military leaders to combine the political power which society thrusts upon them with the military professionalism without which society cannot endure".[26]

[26] Huntington, *op.cit.*, p. 464

3. Military ties with business and industrial elites

Although the professional military man prefers traditional order and social stability to social change, it would be a mistake to infer that military conservatism applies equally strongly to *technological* change. On the contrary, one may with greater justification propose the thesis that the military profession from its beginning has been closely associated with technological innovations. In military education, as we have seen, technical schools were often established well before schools for the infantry and cavalry. In the United States, West Point was early in the forefront of technical education.[27]

[27] See above, ch. 1, sec. 6.

This thesis is not without qualification, however. A number of examples of resistance by military men to military inventions do exist, particularly with regard to the period of the first World War. In 1915 the French General Headquarters turned down a proposal of development of the tank weapon characterizing the new vehicles as "engines not susceptible of lending themselves to any military use". In England, Haig and Kitchener only reluctantly accepted the machine gun, Haig calling it "a much overrated weapon". The equipment of the English forces with the machine gun was to a large extent the effect of Lloyd George's concern about the waste of men being mowed down by the Germans who had been more quick in adopting the new weapon.[28] Military men on the Allied side in the war were less prone to advocate the development of heavy artillery, which contributed to the superiority of the Central Powers in this field during the beginning of the war.[29] French politicians in 1910 pointed to the military use of aviation, getting the answer from Foch, Director of the Ecole de Guerre, that "for the army, the aeroplane is zero".[30]

It is questionable, however, how far one can generalize from these cases. First, it seems that the resistance to weapons innovations often came from the older generation of officers while the younger took active part in the development of new weapons. For instance, while a few English generals were reluctant to adopt the tank as a weapon, junior officers played an important role in its development and promotion for war use.[31] Second, the resistance to new weapons by English military leaders may have been largely caused by a *deformation professionel* resulting from their experiences in the Boer War, in which horsemanship had counted for more than trench warfare, barbed wire, and heavy artillery. As Lloyd George writes in his memoirs, when the First World War broke out the English generals had the most important lessons of their art to learn.

> Their brains were cluttered with useless lumber, packed in every niche and corner.... For instance, take their ridiculous cavalry obsession. In a war where heavy artillery and engineering and trench work were more in demand than in any war in history we were led by soldiers trained in the cavalry. Haig was persuaded to the end of the war that a time would come when his troopers would one day charge through the gap made by his artillery and convert the German defeat into a headlong scamper for the Rhine. Needless to say, that chance never came.... The Generals themselves were at least fourfifths amateurs, hampered by the wrong training.[32]

Third, there was no marked conservatism exhibited against some other new means of warfare, most importantly the U-boat and the poison gas which, in the judgment of Vagts, the Germans employed "prematurely and too experimentally".[33] And after all, perhaps the most epochmaking innovation of them all, the aeroplane, was making rapid headway. In France, the machine so scornfully turned down by Foch in 1910 was manufactured in no less than 50.000 units during the period of 1915–18; and at the armistice in 1918, the French plane factories employed 186.000 workers. Similarly, during the war Germany produced 48.000 planes in 33 factories.[34]

The situation after the Second World War has been characterized by continuous technological experimentation, witnessed in the development of systems like the Nike-Zeus, Nike-X, Sentinel (all three anti-ballistic missile systems), the Fractional Orbit Bombarding System (FOBS), the Ballistic Missile Early Warning System (BMEWS), Semi-

[28] Vagts, *op.cit.*, p. 233
[29] Ibid., p. 372
[30] Ibid., p. 372–373
[31] Ibid., p. 232
[32] David Lloyd George, *War Memoirs*, Boston 1933–37, vol. VI, pp. 338–347
[33] Vagts, *op.cit.*, p. 231
[34] J. Boudet et al. (eds.), *Armeérnas Världshistoria*, vol. IV, Stockholm, 1969: AB Svensk Litteratur, table 1.

Automatic Ground Environment air-defense system (SAGE), Multiple Independently Targetable Re-entry Vehicles (MIRV), missiles like the Polaris, Poseidon, Minuteman I, II, and III, and Titan, and aircraft such as the nuclear-powered plane (development cost $ 1.025 billion), the B-70 bomber ($ 1.5 billion), and the Seamaster jet-powered seaplane ($ 361 million), the three latter now being judged as "unfeasible and wasteful" development projects.[35]

With the increasing emphasis on technological development, the contacts and personnel alliances between the military and industrial establishments have tended to widen. In March, 1969, Senator William Proxmire released figures showing that the 100 most prominent defense contractors in the United States employ "some 2.072 retired military officers of the rank of colonel or Navy captain or above" (compared to 721 in 1959).[36] The Sentinel ABM system alone has been estimated to involve more than 15.000 companies for its development and production.[37] The impact of such developments has caused the American debate on the military-industrial complex to change perspective. At the publication of Mills's *The Power Elite* the argument to a large extent was about whether the military-industrial complex *existed* or not; today, one more commonly finds arguments *for* or *against* it, and the question of its existence seems to be definitely answered in the affirmative.[38]

The far-reaching coincidence of military and industrial interests may be expected to cause a modification and "modernization" of the traditional military conservatism. It should also be pointed out that the tremendous growth of logistical functions especially during the present century has made the military manager responsive to economic planning. He is now part of the administration of what is almost a replica of civilian society (but for certain production and reproduction functions). Hence it seems only natural that, as Janowitz has expressed it, traditional conservative identifications should become "compatible with a belief in the need for continuous and decisive governmental intervention in the economic order", especially of course if such intervention is advantageous for the military establishment. "The military approach to the economic system centers on the issue of the military budget, on what, theoretically, the nation needs and can afford for national security."[39]

Today, to a large extent, major economic power-groups have replaced the old aristocratic elites as the primary supporters of the military system, and we may expect military conservatism to exhibit a more up-to-date profile as it becomes modeled upon the political outlooks of business and industry. However, the new outlooks do not necessarily guarantee a more liberal attitude to major social reforms since, as Janowitz emphasizes, the military's views of the economic system are primarily motivated by professional self-interest.

---

[35] Adam Yarmolinsky, "The Problem of Momentum", in Abram Chayes and Jerome B. Wiesner (eds.), *ABM – An Evaluation of the Decision to Deploy an Antiballistic Missile System*, New York, 1969: Signet Books, p. 146

[36] *Ibid.*, p. 148

[37] *Ibid.*, p. 146

[38] The *Chicago Daily News*, April 16, 1969, carried the following news announcement:

Barry: 'Thank heaven'

Washington (UPI) – Sen. Barry M. Goldwater (R–Ariz.) said Tuesday that the United States "should thank heavens" for the military-industrial complex as "the bubble under which our nation thrives and prospers".

"Its ultimate aim is peace in our time regardless of the aggressive, militaristic image that the left wing is attempting to give it", the 1964 GOP presidential candidate said in a Senate speech.

[39] Morris Janowitz, *The Professional Soldier*, p. 246

Summary and conclusions

1. Military conservatism in part reflects dominant elements in "classical" conservatism. It tends to emphasize order, hierarchy, and the "stabilizing" institutions of society (church, family, private poperty). It maintains a pessimistic image of human nature and is dubious of the prospects of avoiding or eliminating war, such pessimism facilitating adjustment to the military professional role. Conversely, an optimistic image of man, optimism about the possibilities of avoiding war, and radical values are at variance with successful military professional performance.

2. The military's historical associations with ruling elites have tended to support and reinforce conservative values, with the exception of cases where armies have recent revolutionary experiences. Conservatism and professionalization may be assumed to be mutually supportive.

3. A third and more recent element in military conservatism stems from its ties with modern industrial and technological elites, leading to an acceptance of rapid technological developments. This does not, however, necessarily imply a more progressive attitude to major social reforms.

4. Managerial training and professional experiences in running the highly complex military establishment will increase the confidence of military leaders that they are competent to also administer the civilian sphere (since the military replicates a large number of functions found in civilian society). To the extent that they command enough power to bring about this transfer of skills and to intervene in the political process, military political ideology will adversely affect programs aiming at social equality. It will support technological research and development, particularly if such programs favor the interests of the military establishment. The combination of military conservatism, technologism, and professional expertise represents a formidable force in the political struggle over the utilization and distribution of national economic resources.

ELEVEN

NORMATIVE INFLUENCE AND THE INSTITUTIONALIZATION OF PROFESSIONAL VALUES

An examination of the social power of the military profession requires the consideration of two types of processes: one relatively long-term, diffuse, and unconspicuous, the other (since it deals with specific issues) more limited in time, overt and dramatic.

The first is the military's *normative influence*, i.e., its ability to affect the diffusion of certain *values*, relevant to the profession's issue area, among the public, as well as the long-range capacity to adhere to these values. Normative influence is dependent on the charismatic properties of the profession, i.e., its ability to "restore emotion, awe, and magic"[1] to the values and to the professional man as their guardian.

The second is the military's *political power*, or the ability of the military profession to overcome resistance in actual decisionmaking, in order to implement military *objectives*. This implies that specific professional objectives do exist, that is, situations arise in which the profession does not act as a subordinate organ of the polity but rather as an independent one. Political power will be discussed in further detail in the following chapter.

The total influence of the military in society cannot be evaluated solely with regard to its overt power in the process of political decisionmaking. As was hypothesized in Part II, the more professionalized a certain group, the more "pure" and coherent its cluster of professional values. As a further hypothesis — to be supported in sections 2 and 3 below — the group will strive to gain followers and supporters of its dominant values among other groups. For example, the prestige and influence of the clergy will be dependent on public acceptance of religious values; the role of physician hinges on public belief in the need for hygiene and medical treatment; and that of the professional officer on values such as nationalism and alarmism (as argued in Part II, chs. 5 and 7).

In other words, the profession will work toward the institutionalization of certain values. To achieve this, organization is a highly valuable instrument. Organization helps in two ways.

First, it can be employed to support an issue with detailed and

---

[1] D.G. MacRae, "Charisma", in Julius Gould and William L Kolb, *A Dictionary of the Social Sciences*, New York 1964: Free Press, p. 84

well-researched arguments, and to back it up with expertise and scientific know-how in order to carry it safely through the political machinery. Secondly, it can be employed to do exactly the opposite, i.e. to bring about that certain "hazardous" issues are "organized out" of the political realm. As Schattschneider has put it, "organization is the mobilization of bias".[2]

For example, the stronger the nationalistic and alarmistic atmosphere in the society, and the more distrust that is generated against other nations, the less likely will it be that civilian "dovish" groups can successfully raise proposals for disarmament and reduction of the military budget. In such an atmosphere, then, certain issues will tend to be eliminated from political consideration, and will not develop into concrete objects of political strife. Thus, to the extent that the military is able to exert influence over social values, this influence will be indirect rather than direct, and may well go unnoticed by the general public and the polity. In short, the military's influence will concern the making of "nondecisions" rather than decisions.

The concept on nondecision-making was introduced some years ago by Bachrach and Baratz in their article "Two faces of power". They argue that power is not only exercised in concrete decisions but also — the "second face" of power — in processes through which a person or a group attempts to restrict actual decisionmaking to relatively "safe" issues. They ask the rhetorical question

> Can a sound concept of power be predicated on the assumption that power is totally embodied and fully reflected in "concrete decisions" or in activity bearing directly upon their making?

and answer

> We think not. Of course power is exercised when A participates in the making of decisions that affect B. But power is also exercised when A devotes his energies to creating or reinforcing practices that limit the scope of the political process to public consideration of only those issues that are comparatively innocuous to A. To the extent that A succeeds in doing this, B is prevented, for all practical purposes, from bringing to the fore any issues that might in their resolution be seriously detrimental to A's set of preferences.[3]

The status of the professional officer as an expert on matters of war and strategy and his integration into an organization where he works full-time as a member of a team with common goals, being able to draw on vast human and material resources, makes him excellently equipped for "organizing out" certain political solutions that are judged unacceptable. The arsenal of military public relations may be activated to neutralize opposition against an ongoing war from certain sectors of the public. The military's connections with patriotic organizations (e.g., veterans) may be utilized for the mobilization of support in local communities. Lobbying and bargaining with political leaders may secure that only "safe" issues are raised.

The importance of organization for the institutionalization of values will be dealt with in further detail below (sec. 3.1), in connection with

---

[2] E.E. Schattschneider, *The Semi-Sovereign People*, New York, 1960, p. 71. Qouted in Peter Bachrach and Morton S. Baratz, "Two Faces of Power", *American Political Science Review*, vol. 56, March, 1952, pp. 947–952

[3] Bachrach and Baratz, *op.cit.*, p. 948

some arguments developed by Arthur L. Stinchcombe. Before doing that, however, it is appropriate to discuss, first, the difference between the concepts of value and objective, and second, some determinants of normative power. The chapter will conclude with some remarks concerning the pluralist theory of elites, and the problem of overlap between strategy and politics.

1. Values and objectives

The difficulties in defining "value" are considerable. Following Robin M. Williams, it is convenient to define it in operational terms as a *direction of interest*. We may test the existence of certain values by observing to what people "pay attention" in their daily activities and speech habits. Values can often be inferred from verbal materials. "In argument, for instance, the statements arousing 'heat', emotion, and so on, are clues to values. In assertion and counterassertion, there frequently emerges level after level of favorable or unfavorable reactions – a 'regress' that in certain cases can be followed back to certain irreducible and ultimate values".[3b] Values prominent in the military professional issue area are, as pointed out in Part II, nationalism, pessimism about the nature of man, and alarmism. (Although conservatism and authoritarianism are consonant with these dimensions, it is appropriate to view them as depending on other major variables rather than as independent dimensions). To the extent that the military is able to enforce and support these values among the general public, it has normative influence. We expect this to be visible in, for instance, the degree to which the role of officer is positively appreciated (cathected) among the population.

By "objective" is meant whatever is (perceived as) profitable to the professional group, in the sense of its being an instrument for satisfying professional interests (or the interests of a particular subgroup, such as a military service branch or a junta of officers). Objectives are more concrete, specific, and detailed than values. Alarmism is a military value; to get a particular ground-to-air missile XB 207 Mark II as a defense against an anticipated extension of enemy air power is an objective. Nationalism is a value; to block political attempts at signing a Test Ban Treaty is an objective. Objectives are usually related to values, in the sense that the professional group may regard the implementation of certain objectives as important steps toward fulfilling a certain value. In particularly heated debates one may occasionally find that a value is identified with only one specific objective, i.e., the value can alledgedly be preserved if and only if that objective is achieved (the "if we don't get the bomb, that's the end of democracy" argument).

Objectives are pursued and supported through various techniques, such as lobbying, bargaining with political authorities, or through means of coercion like strikes, boycotts, blackmail, military coups etc. Obviously, the greater the profession's normative influence, the greater will be its

[3b] Robin M. Williams, *American Society*, New York, 1951: Knopf, p. 378

ability to mobilize support for particular objectives. For this reason, power-enhancing strategies will usually be combinations of long-term public relations efforts and specific, well-prepared persuasion attempts aimed at political power-holders.

Values and objectives will also be taken up for further consideration in the next chapter. As I will argue, they are essential elements in military power by prompting the mobilization of resources.

2. Determinants of normative influence

The profession will be able to increase its normative influence as a function of (1) its public relations and propaganda, (2) its performance during earlier historical periods, and (3) the relative presence or absence of crisis situations pertaining to the profession's issue area. Thus, *to the degree* that its public relations are effective, its previous performance successful, and crisis situations are present, the profession will find its normative influence enhanced.

2.1. Public relations and propaganda. Most military establishment keep sub-organizations for public relations. In the United States, the military use of such techniques grew markedly especially during and after World War II. Publicity staffs were instituted in the European and Far Eastern war theaters, as well as in the office of the Chief of Staff. The Pentagon established a multi-branch public relations apparatus, including, among other things,

- a radio and television department (preparing scripts, recordings, pictures)
- a pictorial branch (planning and producing still and motion pictures; providing military films for civilian groups cooperating with the services)
- a magazine and book branch (to maintain liaison with magazine and book publishers, editors, staff writers, and freelance correspondents)
- a speakers and public appearances branch
- a national organizations branch (to maintain liaison with national organizations interested in matters pertaining to the National Military Establishment)
- an analysis branch (to analyze and summarize news and to provide military personnel with information concerning public opinion), and
- an industrial relations branch.[4]

Within the Swedish military establishment, press officers are assigned to local regiments, and the Defense Staff operates a central press and public relations division. Besides, military officers sometimes take active part in information on defense issues in primary and secondary schools. Similar examples could be cited for a large number of other countries.

[4] See Fred J. Cook, *The Warfare State*, New York 1962: Macmillan, pp. 94–95

Thus, through public education, information to national organizations, newspapers, magazines and other media, through public appearances on radio and television, etc., the military may influence the public image of itself and the values it embodies.

2.1.1 Military service as a mediator of patriotic values

Although the military profession does not entertain client-practitioner relationships in the same sense as, for instance, the medical profession, it is nevertheless in a favorable position for influencing those of the male population who have to do military service (whether as draftees of conscripts). Military service can be assumed (a) to implant definitions as to who is the primary potential enemy (and who the primary ally), (b) to create and sustain favorable attitudes to military methods as the chief means of security policy, and in general (c) to support attitudes congenial to the military mind (as defined earlier, chs. 4–8) rather than their opposites.

This issue is also related to the question of the impact of war veterans in the formation and keeping alive of patriotic and martial sentiments among the population.

In the United States today, it has been estimated that the number of living veterans from World War II is 14.9 million, and from the Korean War about 5.7 million. Subtracting from the latter figure those who served also in World War II, the total number of veterans from these two wars add up to about 19.5 million men, or about 30 per cent of the total U.S. male population of 15 years and older.[5] The Vietnam War will raise these figures still higher.

Organizations of war veterans in the United States have a total membership exceeding 4 million. It seems a reasonable hypothesis that these organizations are among the most important opinion leaders for the nurturing of patriotic attitudes in the American society. As former Marine Corps General David M. Shoup has pointed out, the veterans "generally favor military solutions to world problems in the pattern of their own earlier experience, and often assert that their military service and sacrifice should be repeated by the younger generations".[6]

The social heritage of attitudes toward war and toward the military establishment should be singled out as one major aspect of sociological research on armed forces and society. How are the images of war changed over time for given individuals? How is the social learning of such images affected by the existence of large numbers of veterans? Do children of war veterans have different perceptions of wars than children brought up in homes of non-veterans? Although outside the scope of the present study, it is remarkable — in view of the relative ease of getting access to

[5] War veterans from Korean conflict only: 4.568.000. Veterans from World War II: 14.916.000. Male population 15 years and older: 66.324.000. Source: *U.S. Handbook of Facts, Statistics, and Information*, 1968, tables 6 and 387.

[6] David M. Shoup, "The New American Militarism" *The Atlantic*, April 1969.

such data — that this area of attitude research is largely untouched by sociologists and social psychologists.

2.2. Historical performance. The prestige of a particular professional group in a particular society and the extent to which the profession will be able to propagate its values will partly depend on its success or failure during earlier periods of the nation's history. A tradition of successful military engagements will enhance the positive perceptions of the military among the domestic population (cf. Israel in the post-World War II period) and contribute to its status as a symbol of national and patriotic sentiments. A historically unsuccessful military establishment will be at a corresponding disadvantage, and will have to revert either to conscious efforts of re-writing history (e.g., through "stab-in-the-back" legends) or try to compensate its low normative power through gains in political power by various means of political intervention.

2.3. Crisis situations. In general, there will be a tendency for the normative power of a profession to rise when the society (or a major part of it) is threatened by crises relating to the profession's issue area. Crises will serve to bring the role of the professional to increased public attention; the subject matter of the profession (war and international conflict in the military case; major epidemics, in the medical case; economic crises, in the case of economists and business analysts) will get wide coverage in the press and other mass media; as a consequence, actions and recommendations by representatives of the profession will get added weight. The relative increment of normative influence as a result of crises will be larger for a profession such as the military, which is usually very low in prestige during peacetime (see Part II, ch. 9), than for a normally high-prestige occupational group such as physicians.

Crisis periods are often accompanied by requests to political authorities from professional groups to increase the economic and material resources of those groups. Incidents of international conflict tend to be followed by military demands for increased armaments. For example, during the Czech crisis in August, 1968, the Swedish Commander-in-Chief requested that the government buy more war materiel in order to strengthen military preparedness.[7] Military professionals in most countries keep the international tension level under close observation in order to react swiftly and with the "appropriate" resources. Quite aside from their constitutional role as leaders of the military defense, military professionals have a number of other reasons for emphasizing the danger to the nation; expansion of the military budget means better chances for promotion, more interesting jobs, heightened public attention and rising prestige. Because of the tendency of the profession to utilize international conflict incidents to improve military appropriations, even relatively isolated international conflicts (with little probability of spreading) will serve as pump-primers for the military economy in other nations as well.

[7] Åke Ortmark, *De okända makthavarna*, Stockholm 1970: Wahlström & Widstrand, p. 229

The expansion and contraction of military appropriations under the influence of major international crises may be illustrated by the fluctuations in the length of conscription time in Sweden during the nineteenth and twentieth centuries (diagram III). The diagram shows the development in the number of days of first service for the main part of Swedish conscripts, starting at a low 12 days during the earlier part of the nineteenth century, increasing somewhat under the impact of the Crimean War 1853–56, then rising steeply during the 1890's and the first decade of the present century.

This rise seems to be a reflection of the general trend of armament which was common in Europe around the turn of the nineteenth century. The following figures are the outlays for defense in pre-1914 Europe (in millions of pounds).

	1858	1883	1908	1913
Great Britain	23	28	59	77
France	19	31	44	82
Germany	5	20	59	100
Italy	2	12	18	29
Austria-Hungary	11	13	21	24
Russia	19	36	60	92
Other states	16	23	38	82
Total	95	163	299	486
Average price level (1913 = 100)	110	95	90	100
European population, in Millions	278	335	436	452[8]

[8] After Vagts, *A History of Militarism*, p. 333

During the 1900's, the impact of the great wars is clearly visible: conscription time rose to 340 days during World War I and to 450 days in World War II, after having declined sharply in the inter-war period. Again in 1952, the number of days of first service went up from 270 to 304 as a consequence of the Korean War.

A certain part of the peacetime activities of the military is devoted to the task of emphasizing national insecurity and the need to strengthen military preparedness. With the growth of military technology, however, the professional need for alarmist indoctrination may be assumed to decrease somewhat, because the size, cost, and complexity of modern weapons systems automatically secures a considerable degree of continuity in military spending. Before World War II, wars tended to be followed by de-mobilization and national reductions of armaments (although usually not all the way down to the level at which they were before the war). In the post-war era, however, nuclear, missile, and aviation technology has created an almost entirely new situation, with long-term planning and production periods.[9] Today the military is able to point to the difficulties and shortcomings of a "trend-adjusted" defense; and may argue that, because of the long procurement periods, "occasional shifts in the military-political situation must not be taken as a basis for revisions of the defense policy".[10] If military planning must be carried out for conditions that, presumably, will prevail ten to fifteen years ahead, very large proportions of the defense budget will have to

[9] Cf. John Kenneth Galbraith, *The New Industrial State*, Boston 1967, Houghton-Mifflin, pp. 310–313

[10] Commander Bengt Lundvall, public speech, Uppsala university, November 16, 1965.

DIAGRAM III

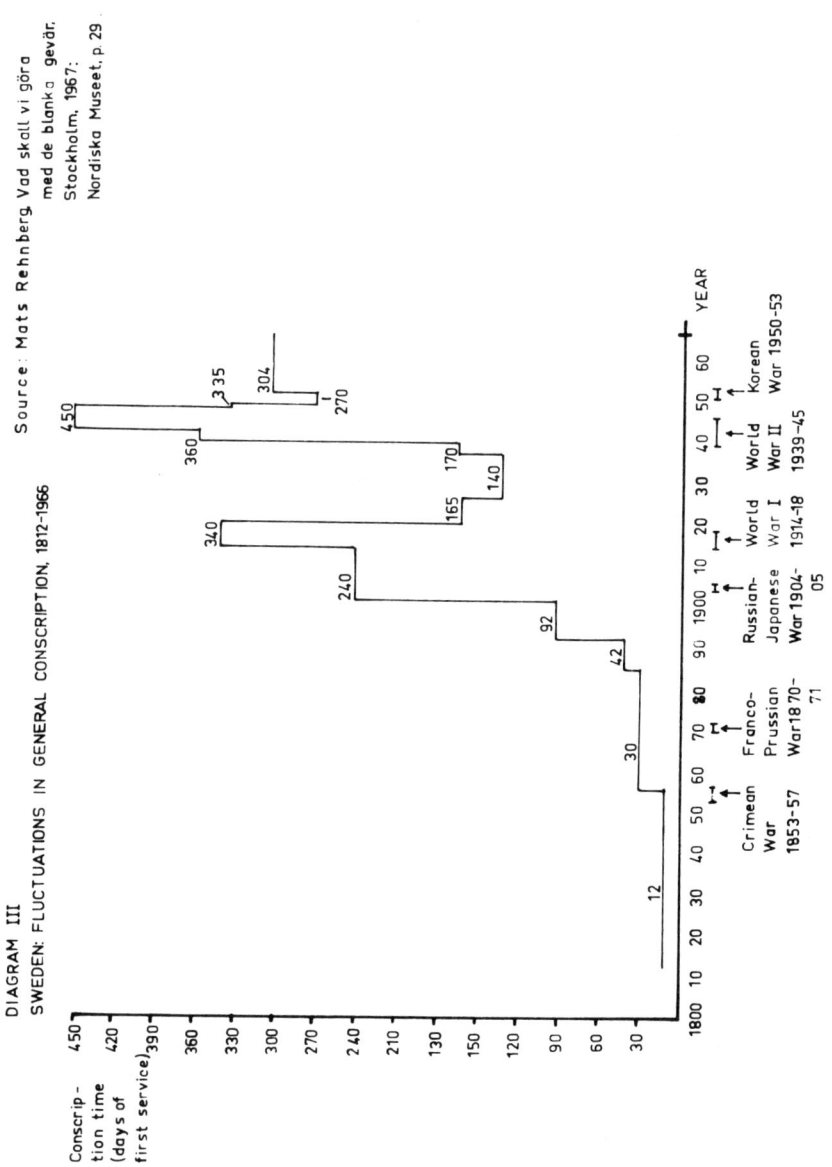

remain fixed for long time. This budget stability will constitute a partial compensation for losses in normative power "if peace breaks out". For example, it has been argued that the Safeguard ABM system will guarantee a certain number of military officers, military technicians, operational staffs and supply personnel continued employment during the 1970's.[11]

3. Intra-professional mechanisms for the institutionalization of values

To be a successful instrument for the propagation of major social values, the profession has to start by institutionalizing those values among its own members, in order to turn them into the firm believers that form the core of the professional group, or, as Everett Hughes put it, into persons who "profess to know better than others the nature of certain matters".

The normative power of the profession is positively related to the degree of conviction among the members that they, and they alone, possess *the* theoretical and practical knowledge necessary for "appropriate" solutions in the issue area. To create the core group of convinced members, professions see to it that persons with congenial rather than uncongenial attitudes are selected and promoted, and that the members receive the proper socialization.

This has already been discussed earlier (chs. 3 and 4). However, it should be pointed out in this connection that the process of institutionalization of values, as it has been described by Arthur L. Stinchcombe, involves precisely those mechanisms that here have been treated as major stages in professionalization$_2$. For, as Stinchcombe points out, "the key to institutionalization /of a value/ is to concentrate power in the hands of those who believe in that value".[12] To achieve their goals, institutions utilize (a) control over selection of members, (b) control over socialization of members, (c) control over the conditions of incumbency in powerful roles (for example, by "constitution-writing"), and (d) the means of letting powerful members of the institution "serve as ego-ideals for ambitious young men".[13] The parallel between institutionalization and professionalization should be readily apparent.

Like most power-holders, professions are usually anxious not to emphasize their own potential social influence and, as a consequence, prefer to create and support the image of themselves as public service-oriented bodies, following but not leading public opinion, and highly responsive to the wishes of their clients.[14] As Maury Feld has noted with regard to the American military, "the civilian concept of military professionalism is likely to be drawn on the model of scientific objectivity and moral neutrality".[15]

3.1. The formal organization as a resource for the institutionalization of values.

Some professions, for example the military, the clergy, and medicine, are able to draw upon a highly useful device for effective dealings with

[11] A similar example of saving-the-situation strategy is provided by the efforts of the American military to persuade political leaders and the public after the end of World War II that a system of universal military training (UMT) was necessary to counter Communist aggression. See Fred J. Cook, *The Warfare State*, pp. 98–105, and above, Ch. 7, footnote 1b. Some main arguments in the ABM defense debate are reviewed in the following articles: Jerome B. Wiesner and Herbert F. York, "National Security and the Nuclear Test Ban", *Scientific American*, October 1964; Robert S. McNamara, "A 'Light' ABM system", *Scientific American*, September 1967; Richard L. Garwin and Hans A. Bethe, "Anti-Ballistic Missile Systems", *Scientific American*, March 1968. All reprinted in *Science, Conflict, and Society – Readings from Scientific American*, San Francisco, 1970: W.H. Freeman & Co. (with introductions by Garret Hardin). A good overview is given by Abram Chayes et al. in Abram Chayes and Jerome B. Wiesner, *ABM – An Evaluation of the Decision to Deploy and Antiballistic Missile System*, New York 1969: Signet Books. This book contains a large number of papers devoted to various aspects of the ABM system: technological, strategic, and arm control. See also *Anti-Ballistic Missile: Yes or No? A Special Report from the Center for the Study of Democratic Institutions*, New York 1968: Hill and Wang

clients, the general public, and political powerholders:[16] the formal organization.

The formal organization is an instrument for the long-range planning of professional affairs; for the rational collection of data relevant to professional practices; for public relations and lobbying work in support of the profession's values and objectives; for the management of institutions of professional education, and for updating their curricula to follow the modifications of professional theory.

The formal organization is of importance to the effectiveness of professional work because of its (a) *technological resources* (office machines, means of transportation, printing equipment, telecommunication materiel, computers, weapons), (b) *spatial base* (offices, classrooms, barracks, clubs, shops, canteens) and (c) *assistant staff*, usually distinguished from the professional members by means of insignia, spatial arrangements, and relative absence of status symbols (orderlies, secretaries, chauffeurs, mailmen, punch operators, etc).

The formal organization provides a structured framework for persons engaged in giving full-time attention to certain professional values and objectives, and is highly instrumental in maintaining professional power both vis-a-vis the organs of the state and the public. As Stinchcombe has appositely put it, values backed by the rationality of organization have definite advantages in competition with alternative values.

Directly or indirectly, power-holders in institutions get paid for thinking about how to achieve and preserve the values and interests embodied in an institution. The more elaborate an argument in favor of value, the more extensive the data collection on which a solution to a problem is based, the more explicitly alternatives are explored and evaluated, and the longer the time span planned for, the more likely it is that the analysis was done by somebody who gets paid for it.[17]

The military formal organization is a set-up for the planning of and preoccupation with methods of armed warfare which, partly as an effect of the impact of military organizations, have become institutionalized as the "established" methods of territorial defense and expansion; and, as a consequence, the "values and interests embodied" in the military institution are at variancee withother less "established" methods of promoting national security (such as international agreements, arms limitation, and academic research on the social causes of conflicts).

In Alfred Vagts's historical survey of attempts at international disarmament, the military figure as systematic opponents of such plans. Summing up the experiences from the Hague conferences in 1898-99 and 1907, as well as later disarmament negotiations, he states: "The active opponents of actual restrictions were always soldiers or sailors who wanted armament or rearmament. They were never too fastidious in their argumentation, like the French general who in Washington /at the disarmament conference in 1921/ defended the continued use of submarines by saying: "You could not stop the progress of humanity!" "[18] And at the first Hague conference, military men from the participating countries got together in defense of their corporate interests in disarmament: "In

[12] Arthur L. Stinchcombe, *Constructing Social Theories,* New York, 1968: Harcourt, Brace, and World, p. 108.

[13] *Ibid.,* pp. 107–117.

[14] Cf. below, sec. 4.

[15] Maury D. Feld, "Military Self-Image in a Technological Environment", in Morris Janowitz (ed.), *The New Military,* New York, 1964: Russell Sage Foundation.

[16] Because of its significance for the profession's political strategies, the section on formal organizations is related also to a number of topics to be covered in the following chapter.

[17] Stinchcombe, *op.cit.*, p. 114.

[18] Vagts, *A History of Militarism,* p. 403.

their zeal they often forgot their national divisions and like members of a professional *internationale* or fraternity united in agreeing to yield as little as possible to the civilians".[19]

The emergence of the military profession and the growth of military organizations has meant the development of a "war establishment", committed to military armament as a means for preventing conflicts and to armed warfare for solving them once they erupt. The "war establishment" is permanently active, commands huge economic resources, is backed by the calculated rationality of formal organizations, and is able to mobilize support for its values and objectives among its own members as well as among civilian groups with nationalistic orientations. Efforts to employ non-military solutions for the reduction of international tension are usually less well and less permanently organized (as exemplified by the many ad hoc disarmament conferences); they are supported neither by formal organizations nor by economic resources comparable in size and scope with those of the military and, as a consequence, have not been particularly successful in mobilizing public support for disarmament and collective security plans. However, the significance of the UN and other similar organizations (as well as the various institutes for conflict research, e.g., the Peace Research Institute in Oslo (PRIO) and the Stockholm Peace Research Institute (SIPRI) should not be judged solely on the basis of their concrete achievements in the solution of specific conflicts, but also against the fact that they represent *functional equivalents* of the traditional military solutions to national security; that is, part of their output consists of the institutionalization of *other values* than those of the military. And this is important since, in a wide context, civilian control of the military involves not only constitutional checks and balances, but also the conscious, long-term, and well-funded support of groups and individuals working for alternative goals.

[19] *Ibid.*, p. 400.

Two digressions

4. Elite competition vs. elite autonomy

Pluralist theory (as represented by Riesman,[20] Dahl,[21] Keller [22] and others) has been bent on assuming that the simultaneous existence of several elites implies a *political balance of interests* somewhat along the lines Karl Mannheim seems to have had in mind when he stated that the elites end up by cancelling each other out.[23] However, this is a dubious implication, since a power balance to the benefit of the public need not exist unless the elites compete with each other in the same issue area; and this is rarely the case. These arguments have recently been developed in an article by Newton.

[20] David Riesman, *The Lonely Crowd,* New York, 1953: Doubleday.

[21] Robert A. Dahl, *Who Governs?*, New Haven, 1961: Yale University Press.

[22] Suzanne Keller, *Beyond the Ruling Class,* New York, 1968: Random House.

[23] *Man and Society in an Age of Reconstruction,* London, 1968, pp. 86 ff.

De Tocqueville pointed to the role of secondary associations in maintaining democratic stability in the United States. Lord Acton stated that "Liberty is possible only in a society where there are centers of organization other than the political". Within this line of argument it is only a short step to saying that a pluralist democracy is a political system in which political power is divided among different elites. As Arnold M. Rose states/in *The Power Structure/*, "Pluralism is a theory of the power structure in which power is conceived of as dispersed, and different elites are dominant in different issue areas". This may be an adequate definition of pluralism, but it is not an adequate definition of pluralist democracy because *it is still possible,* within the terms of Rose's definition, *for each elite to dominate one particular area of political activity.*[24]

[24] K. Newton, "A Critique of the Pluralist Model", *Acta Sociologica*, 1969, vol, 12, no. 4, p. 211 (italics added).

The effective-competition model implicit in the pluralist position is inadequate as a description of the political behavior of elites and professional groups. One may ask: where are the mechanisms "cancelling out" the power of the Pentagon against that of the American Medical Association, or of the latter against the American Bar Association? What evidence and, indeed, what motivations, exist for a competition between the AMA and the ABA? Between the military and industrial interest groups? In general, these groups have very little to win by interfering in each other's affairs; on the contrary, they have much to win by *not* interfering, since lawyers sometimes need the support of doctors, doctors need lawyers, the military is dependent upon industry, and industry on the military.

In *Beyond the Ruling Class,* Suzanne Keller discusses the emergence of "strategic elites" and notes, in true pluralist spirit, that elite specialization imposes an important amount of equality: "Today no single strategic elite has absolute power or priority, none can hold power forever, and none determines the pattern of selection and recruitment for the rest."[25] While this may be an important condition for the freedom of *despotism,* it is obviously inadequate for insuring *democracy,* since the elites may, because of their superior organization, defend their interests against claims to social equality from the unprivileged strata. Elite autonomy does not guarantee elite *responsiveness* to the needs of the mass of the people. As Bottomore has pointed out.

[25] Suzanne Keller, *op.cit.* pp. 277–278.

In the Western societies the elites stand, for the most part, on one side of the great barrier constituted by class divisions; and so an entirely misleading view of political life is created if we concentrate our attention upon the competition among the elites, and fail to examine the conflicts between classes and the ways in which elites are connected with the various social classes.[26]

[26] T.B. Bottomore, *Elites and Society,* New York, 1964: Basic Books, p. 118.

That Keller actually means something more than simply freedom from despotism is obvious from her definition of elites: "Elites are effective and responsible minorities – effective as regards the performance of activities of interest and concern to others to whom these elites are responsive".[27] This definition begs the most important question as to *how* responsible a certain elite is to the interest of those it is supposed to serve. The responsiveness of elites is an empirical, not a definitional problem. Such circumstances as the opposition of the AMA to general health insurance, the lobbying efforts of the military to further professional objectives, and the resistance of voluntary welfare organizations to state and federal social security programs in the United States,[28]

[27] Keller, *op.cit.*, p. 4.

[28] See Roy Lubove, *The Struggle for Social Security 1900 – 1935,* Cambridge, Massachusetts, 1968: Harvard University Press.

seem to point at the fact that elites often work for their own interests rather than those of major popular strata. Class differences may be upheld and permanented even though the *elites* have considerable autonomy and freedom.

The conception of the "cancelling out" effects of a plurality of social elites may be represented by path (1) in diagram IV. The conflicting view of autonomous elite power is represented by path (2), for convenience labeled the differentiated power elite model. The degree to which scheme (2) fulfills important criteria of democracy (such as public influence on major decisions, access to decisionmakers, social equality, etc) is dependent on the responsiveness of the elites – and, as pointed out above, this has to be asserted by research on the behavior of each individual elite. As already implied, in order to "cancel out" the power of the military institution there have to exist other institutions *representing competing values in the same issue area.* This is hardly the case today; it would be possible to enhance the truth value of the pluralist model in a sort of backward way by deliberately *creating* such institutions (or to strengthen those embryonic attempts that already exist) and thereby bring about the effective competition situation assumed in model (1). (For an example of a proposed solution, see ch. 13, sec. 2.1.)

DIAGRAM IV

The pluralist and differentiated power elite models

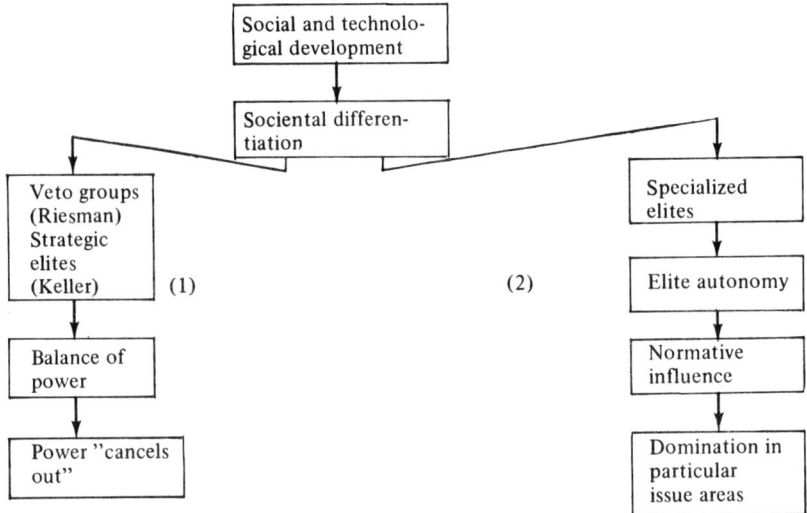

5. The overlap between strategy and politics

As we have seen, some pluralist theoreticians claim that the differentiation and specialization of authority constricts the power of social elites. At least in one case (Suzanne Keller) this notion is upheld with regard to the military by contending that political and military affairs may be clearly separated: "where the politician may tread, the general may not",[29] i.e., the authority of politicians counterbalances the power of generals. Keller's statement implies that the jurisdiction of each group is sufficiently clearly defined, so that the politician will be able to determine when military experts transgress the boundaries of their proper field.

If this contention were correct, there should be no problem in allocating each one of the following decisions either to the political or the military field. As a matter of fact, however, they are all examples of areas where both generals and politicians have left their footprints, the officers often claiming the issue to be mainly or solely military.

(1) Starting unlimited submarine warfare (Germany, World War I)
(2) Starting the development of atomic weapons (the Manhattan project, USA, World War II)
(3) Keeping Algeria French or not (France, late 1950's and early 1960's)
(4) Crossing the Yalu River (United States, Korean War)
(5) Starting the deployment of anti-ballistic missiles (USA, various approaches since the mid-1950's)[30]

To find the demarcation-line between strategy and politics is and old dream of military men. For instance, the older Moltke was time and again irritated by the disruption of his calculations by "unpleasant political realities", and maintained that when the war has started, it is strictly military territory; at the end of it, the politician may step in again. Politics should operate "decisively at the beginning and the end /of the conflict/, of course in such a manner that it refrains from increasing its demands during the war's duration or from being satisfied with an inadequate success."[31] Strategy, he argued, aids politics best "working only for /political/ objectives, but in its operations independent of /politics/".[32]

The overlap of strategy and politics constitutes a constant potential source of strain in civil-military relations. To solve these contradictions, the theory of civil-military relations in the democratic state prescribes that the political authorities reign supreme in setting the goals for the utilization of military forces. In practice, however, politicians are forced at some level in the hierarchy of decisions to leave the responsibility to military commanders. This level can rarely be determined on beforehand, and therefore the political influence in strategy tends to be determined in an *ad hoc* fashion.

During the Khe Sanh affair, the American Joint Chiefs of Staff were requested to sign an assurance that the outpost "can and should be defended". Although the latter part of this statement can very well be regarded as a political and not a military judgment, the Chief Executive in this instance chose to leave the decision to the military.[33] The higher the frequency of such instances, the greater the incidence of what Janowitz has termed "unanticipated militarism", i.e., militarism develo-

[29] Keller, *op.cit.*, p. 277.

[30] Some further arguments along these lines may be found in Ronald J. Stupak, "The Military's Ideological Challenge to Civilian Authority in Post-World War II France", *Orbis,* 1968, Summer, pp. 582-604. Stupak takes the debate on the ratification of the Test Ban Treaty in the United States in 1963 as an example of the ambiguity as to what is a "strategic" as against a "political" issue. "It was extremely difficult to separate the treaty's political aspects from its military ones. Hence, some military men sounded like politicians, while certain politicians sounded lika military strategists".

[31] General Field Marshal Helmuth Count von Moltke, "Über Strategie", *Kriegsgeschichtliche Einzelschriften,* xiii, I (quoted in Craig, *The Politics of the Prussian Army,* p. 216). Cf. also the argument by General Sherman, ch. 13, sec. 2.1.

[32] *Ibid.,* loc. cit.

[33] Cf. Eugene McCarthy, "The Power of the Pentagon", *Saturday Review* December 21, 1968.

ping from lack of traditions for controlling the military establishment, and "from a failure of civilian leaders to act relevantly and constantly".[34] Obviously, since the number of decisions in the area of strategical and political overlap is very large, to act "relevantly and constantly" demands from politicians a high degree of attention to military issues and very large civilian personnel resources employed in control functions.

To the extent that political authorities wish to avoid military interference in politics, they themselves will have to interfere in strategy. However, with the advent of thermonuclear weapons, the acceleration of military technology, the emergence of world-wide international defense alliances, and the involvement of the military forces of the super-powers in the domestic conflicts of foreign countries, the potential field of civilian control has grown so large and complex that political authorities gradually have been forced to relinquish civilian control and have had to allocate a variety of politically related issues to the military.

As described by Masland and Radway with regard to the United States, this tradition was established as a consequence of the overseas engagements of the US Army in World War II. As the war drew to a close, military officers became involved even in "controversial questions of foreign policy". "By the time of the Potsdam conference, papers were being prepared /by officers in the Operations Division of the War Department General Staff/ on such frankly political topics as Soviet intentions with respect to expansion, American policy toward Indo-China, and the terms of Japanese surrender".[35] The first great task immediately after the war was the interim administration of foreign countries such as Germany, Japan, Austria, Trieste, and many Pacific islands.

The main burden here was assumed by the Army. Between 1945 and 1952 General MacArthur's headquarters labored to restore Japan to the family of nations. Military officers helped to revise the constitution and laws. They participated in determining the level and structure of industry; in reforming the monetary and financial system; in guiding the trade union movement and political parties; and in supervising the press and the educational system. Finally, they played an important role in the writing of the peace treaty.[36]

Similar tasks were carried out by US Army officers in Germany under the command of General Lucius D. Clay.[37]

The document reprinted in the Appendix of this chapter may be considered as a further and more recent example of the explicit training in the estimation of political questions given to US Army officers. The paper, used for instruction at the US Army Command and General Staff College, Fort Leavenworth, Kansas, contains a methodology for the strategic appraisal of the internal situation in Cambodia, as of 1968.[38] The problem, as stated in the paper, is "to determine if a need exists in Cambodia for the provision of US military assistance under the authorization of the Foreign Assistance Act of 1961". The premises for the discussion include statements about the size and ideology of insurgent forces; the estimated time before the Cambodian government will fall to the insurgents, provided that no military assistance is given;

[34] Janowitz, *The Professional Soldier,* p. 14.

[35] John W. Masland and Laurence I. Radway, *Soldiers and Scholars,* Princeton, New Jersey, 1957: Princeton University Press, p. 15.

[36] *Ibid.,* pp. 16-17.

[37] *Ibid.,* p. 17.

[38] For space considerations a number of appendices to the original document have had to be left out. These include data on the physical environment, economic, political, sociological, and military variables.

I have not been able to establish the relationship between this document and the subsequent (spring 1970) American intervention in Cambodia. Some statements in the paper can be interpreted as anticipating an intervention (e.g., 3 c and 5); but, if so, neither the anticipated time of the year (August), nor the positive appraisal of Prince Sihanouk coincide with later events. Further, no indication is given as to the role of the Lon Nol military faction, later to carry out its coup d'etat. The general tenor of the paper, however, suggests that the author(s) have acted on the assumption of a fairly rapid change in US policy vis-a-vis Cambodia.

Cambodia's national and political objectives as interpreted by US diplomatic personnel; US objectives in the Far East ("to assist free nations in preserving their security"): ,and the purpose of giving US military aid ("to foster US national interests"). On the basis of the premises, it is concluded that "a need does exist in Cambodia for US economic and military assistance and that such a need is urgent".

Strategic training, as illustrated by this document, involves premises relating to (1) the internal conditions of other countries, (2) interpretations of the political intentions and objectives of foregin government, and (3) conclusions as to recommendable decisions by the domestic government. Again, it is obvious that such training cannot avoid but going very explicitly into the political arena.

Governmental decisions in security policy represent choices between alternatives on the basis of data from various sources, among them the military profession. To the extent that the military profession makes policy recommendations — which appears unavoidable given the earlier arguments — the freedom of action of the political authority will be circumscribed by military professional authority, i.e., will be biased in the direction of military interests. To the extent that the public image of military professionalism is "drawn on the model of scientific objectivity and moral neutrality", it badly represents the complicated realities of civil-military interaction and the impact of military role expansion in the field of politics.

Summary and conclusions

1. The impact of the military on social life has been discussed with regard to (a) normative influence, and (b) political power (which will be considered more fully in the next chapter). Normative influence is the ability of the military to affect the diffusion of certain values (e.g., nationalism, alarmism) among the public. Political power is the ability of the military profession to overcome resistance in actual decisionmaking, in order to implement military objectives.

2. As determinants of normative influence we noted, first, public relations and propaganda, second, the historical record of the profession, and third, crisis situations (tending to raise the prestige of and public attention to the role of the profession in social life).

3. To influence the institutionalization of values, the profession first has to secure loyalty to those values among its own members. This is brought about by, among other things, control over selection and socialization of members. We have noted the parallel between professionalization$_2$ and institutionalization.

4. The formal organization provides a framework for persons giving full-time attention to certain professional objectives and values, and constitutes a primary asset for the dissemination and preservation of professional values. The military organization puts a priority on values

such as nationalism and alarmism; as a consequence, attempts at creating disarmament and international collective security arrangements are likely to meet the corporate and organized resistance of the military.

5. The "war establishment" is permanently active, commands considerable economic resources and can mobilize support for its objectives among its members plus – to the extent it commands normative influence – the population at large. Forces working for non-military solutions to international conflicts are less well organized, often active only on an ad hoc basis (i.e., come into existence only on certain occasions), have less economic resources and, as a consequence, have been less successful in mobilizing support for their objectives. The establishment of international organizations and social science-oriented conflict research institutes is significant for the institutionalization of values challenging those of the military profession.

6. Various elites may hold power simultaneously without the power of each being "cancelled out" through the power of others, since each elite may dominate its own particular issue area. Elite autonomy does not guarantee elite responsiveness toward major social groups; for example, it is difficult to find an elite group whose power tends to "cancel out" the power of the military. Civilian control can hardly rely on the market mechanisms implied by pluralist theory *unless* powerful institutions are created that represent competing values in the military profession's traditional competence area.

7. With the growing complexity of modern wapons systems and the prospect that armed conflict may escalate into thermonuclear war, strategic issues more and more take on a political character. The increasingly blurred demarcation line between political and strategic issues tends to create uncertainty as to which military activities can be legitimately controlled by the political authorities. Such issues as the development of atomic weapons, the deployment of anti-ballistic missiles, and strategic appraisals of the need for military assistance to foreign nations seem to present major problems as to the proper jurisdiction of the military. If the arguments advanced earlier concerning the content of military professionalization (p_2) are correct, a widening of that jurisdiction will mean the support of nationalistic and alarmist biases in security policy.

APPENDIX

U.S. ARMY COMMAND AND GENERAL STAFF COLLEGE

Strategic appraisals—a methodology

Section II, Lesson 3. Solutions to First Requirement, Lesson 2, and to First Requirement, Lesson 3

1. This section constitutes a complete, written strategic appraisal of Cambodia. It is not intended to convey that this format is the only appropriate method for conducting a strategic appraisal. It is *a* method, which, in the absence of other direction, will serve to accomplish the purpose.

2. Save this issue for use in Subjects R2312, *US Combat Forces in Internal Defense,* and R1816, *Strategic Appraisals.*

EMBASSY OF THE UNITED STATES OF AMERICA
PHNOM PENH, CAMBODIA
August 197
Strategic Appraisal--Cambodia

1. Problem

To determine if a need exists in Cambodia for the provision of US military assistance under the authorization of the Foreign Assistance Act of 1961.

2. Assumptions

a. Communist forces in Cambodia will continue to receive support from the Peoples' Republic of China and the Democratic Republic of Vietnam.

b. The Cambodian Government will continue to receive support from the majority of the populace.

3. Facts bearing on the problem

a. The United States has, or can obtain, the means to support an assistance program in Cambodia.

b. An insurgent force of about 30,000, currently operating in Cambodia, is Communist-inspired and Communist-led.

c. The latest National Intelligence Estimate concludes that the insurgent force will control Cambodia within about 18 months if the capabilities of the Cambodian Government and its armed forces are not materially improved.

d. The government of Cambodia has requested aid from the United States in the form of military and economic assistance and a MAAG.

e. The Ambassador has directed that efforts be directed at determining Cambodia's national assets and liabilities in relation to her objective of eliminating the Communist threat.

f. US objectives in the Far East are (1) to assist free nations in preserving their security and help them improve their political, economic, and social conditions; and (2) a determination to meet with firmness the Communist-instigated external pressure.

g. The purpose of US military aid is to foster US national interests. Primary among the interests is the security of the United States as threatened by the international Communist movement. Insuring the security of Cambodia will contribute to this aim.

4. Discussion

Cambodia is at a crossroad. Since independence, the government has been making a strenuous drive to achieve an industrialized society and better the lot of the people of this primarily rural nation. All of the contry's assets have been committed to this path, along with whatever "stringless" aid was available. Comparatively remarkable progress was being made until US aid was cut off, albeit at the request of the Cambodian Chief of State. An attempt to replace US aid with grants and loans from Communist countries was marginally successful; but now, with the People's Republic of China and the Democratic Republic of Vietnam apparently backing the Communist insurgents in Cambodia, that source of economic assistance is threatened. Cambodia is virtually in a state of total commitment at this time, attempting unsuccessfully to maintain stability. The insurgents continue to become stronger and their influence in the countryside becomes broader. Cambodia's military forces are not presently capable of insuring the security of the population and destroying the insurgent infrastructure. In fact, they have been hard put to accomplish the former, and this ineffectiveness has aided the Communists by demonstrating the weakness of the government.

It is in the economic and military areas that Cambodia is most weak. Regular military forces can be classified as fair to good, but do not appear to be a match for the idealistically motivated insurgents. Having been employed on nonmilitary tasks for several years, the government forces require intensive training to enable them to engage the insurgents on a near-equal basis. The manpower problem is not acute; however, there is no training base which can administer the necessary training within the time frame required.

The average young Cambodian can be expected to support the

government and rally to the defense of his country, if called. Though loyal when well led, he possesses few of the natural attributes of a soldier, being a product of a primarily rural society based on Buddhist traditions and completely unfamiliar with machinery of any type. His opposite number, the urban youth, is also at a disadvantage, being steeped in traditional French classicism and ill-prepared for the harsh realities of jungle warfare.

Equipment, too, is a problem. The equipment in the hands of the troops is of mixed origin and in a questionable state of readiness. There is little or no equipment available for force expansion. The logistic system is satisfactory for peacetime operations, but is rapidly failing under the pressures of the present situation.

Cambodia's political foundation provides an asset. In theory a democracy, the political structure is sometimes abrogated by the Chief of State for the good of the country. Price Norodom Sihanouk is the hereditary, traditional Prince, the democratically elected Chief of State, and the titular head of the Buddhist hierarchy. He commands the loyalty of a high percentage of his population through his direct appeal to the people on these grounds. Though his rule can, in truth, be described as a paternal dictatorship, it has proved to be stable and progressive. The government is in little danger of overthrow unless the whole country falls to the Communists. The line of succession, in case of Sihanouk's death, appears secured by tradition and acceptable to the people.

5. Conclusion

Considering all aspects of the present situation in Cambodia and her capability to increase efforts toward stability in the face of an increasing, Communist-inspired, externally supported insurgency, it is apparent that Cambodia cannot, without external aid, withstand the present threat and maintain any appreciable degree of government control.

The crisis Cambodia is facing is even more critical when considered in the light of previous experience with the Peoples' Republic of China and the Democratic Republic of Vietnam and their capacity to increase the size and capability of the insurgent forces.

It is, therefore, concluded that a need does exist in Cambodia for US economic and military assistance and that such need is urgent.

6. Recommendation

It is recommended that the National Security Council advise the President of the United States that a need does exist in Cambodia for economic and military aid to enable that country to eliminate Communist insurgency and regain government control of the countryside.

Annexes: A–Draft Action Cable
             B–Discussion: Objectives and Policies
             C–Discussion: Forces and Trends
             D–Discussion: National Style
             E–Discussion: National Power (with five app)

Appendixes: 1–Physical Environment
2–Economics
3–Political
4–Sociological
5–Military

Annex A (Draft Action Cable) to Strategic Appraisal--Cambodia

FROM: US EMBASSY PHNOM PENH
TO: SECSTATE
SECDEF
INFO: DIRCI
DIRDIA
JCS
CSARMY
JOINT STATE-DEFENSE MESSAGE. JCS pass to ISA

Subj: Cambodian Need for US Military/Economic Assistance
1. Conclusion of country team, with which I fully concur, is that a need does exist in Cambodia for US economic and military assistance and that such need is urgent.
2. Recommend NSC advise President that requested aid is needed and that speed is essential.
3. Copies of Strategic Appraisal--Cambodia, written by augmented country team in course of determining above conclusion, are being airmailed to action and info addressees this date.

Note: This draft action cable would be prepared on completion of the strategic appraisal and contains the recommendations of the country team. Placing this message as Annex A provides the approving authority, in this case the Ambassador, with a draft message, ready for his approval, transmitting the recommendations of his country team to the interested offices in Washington.

Annex B (Discussion: Objectives and Policies) to Strategic Appraisal--Cambodia

1. Two opposing objectives and their implementing policies are involved in Cambodia at this time. Cambodia's immediate objective, in the face of the Communist threat, must be to eliminate this menace to security and restore government control throughout the country. Her policy is, first, to resist and attempt to destroy the insurgents; and, second, to request aid from the United States to help accomplish this objective. Security and maintenance of independence have become the country's primary goals in the present situation, though well-being and development remain parts of the national purpose. The goal of social and economic modernization has had to be put aside for the present, in view of more immediate requirements.

2. In direct conflict with Cambodia's plans is the objective of the Communist hierarchy of the Peoples' Republic of China and the Democratic Republic of Vietnam to bring Cambodia under Communist hegemony. The Communist instrument in their goal is active support, with men and materiel, of an attempt to overthrow the Cambodian government through insurgent activity in the countryside.

3. In Cambodia, as in Vietnam earlier, all the nation's assets are committed to the counterinsurgency role, leaving nothing to be applied toward the more vital function of government in an emerging nation state--that of nation building. Priorities must be toward the maintenance of government structure and internal defence of the state; yet, the deep-rooted desires of the people are best fulfilled through more mundane efforts at nation building. The inability of the government to meet all these demands creates an additional area in which the insurgents may operate to garner more support from the people. It is apparently in the light of this twofold nature of the problem that Prince Sihanouk has appealed for both economic and military assistance.

4. This assistance would enable the government to carry out policies of--

a. Increasing the effectiveness of the armed forces.

b. Increasing the effectiveness and acceptability of the national police.

c. Improving the communication system to permit the government to keep the people more informed.

d. Enhancing the ability to establish a secure distribution system to move products from the countryside to district and provincial towns.

e. Providing additional educational and medical facilities with personnel to staff and operate them.

5. Additionally, and important at this time, are US objectives in Southeast Asia--that all nations should be allowed to develop as free and independent countries, secure from outside aggression or subversion, and able to achieve economic, political, and social growth.

6. In Cambodia's case, the Communist and US objectives are in direct opposition, as Communist policy includes external incursions and inhibits the growth of freedom that is both the goal of Cambodia and the United States.

Annex C (Discussion: Forces and Trends) to Strategic Appraisal--Cambodia

1. Predominate forces

There are three predominate forces in Cambodia at this time. They are the drive for social and economic modernization, nationalism, and the conflict of ideologies.

2. Modernization and nationalism

The first of these, the drive for modernization, is exemplified by the emphasis on industrialization and education. The withdrawal of the French and concurrent independence provided the impetus for these developments. The educational system initiated by the French has provided the base for an expanding and improving system. State school enrollments have been increasing yearly and students are being sent abroad by the government for training in such fields as administration, economics, health, engineering, and education. Nationalism, exemplified by a desire to improve Cambodia's image in the international arena, has given impetus to educational improvement by making it a patriotic duty to achieve as much education as possible.

3. Government action

Government attacks on the factors limiting economic growth have been energetic and many-sided. The government has sought and obtained financial and technical aid from all sources willing to provide it. Plans for economic modernization have put emphasis on the expansion of education, as mentioned above, health, communications, transportation, irrigation and flood control, and the development of small private enterprises.

4. Conflict of ideologies

The two forces, economic and social modernization and nationalism, are now being opposed by the conflict of ideologies or, more specifically, communism, in the form of internal conflict sponsored and abetted by the Peoples' Republic of China and the Democratic Republic of Vietnam. Not only is this conflict aimed at undermining and overthrowing the present government of Cambodia, but, more to the point, the limited assets of the country must now be expended toward resisting the incursion and are not available for the advancement of Cambodia's economic viability. The trend of armed Communist incursion must now be considered the primary force, and it is working directly against the objectives of the country.

Annex D (Discussion: National Style) to Strategic Appraisal--Cambodia

1. General

The national style of Cambodia is almost completely embodied in her Chief of State, Prince Norodom Sihanouk. This is not to imply that the actions and reactions of Prince Sihanouk are typical of the Cambodian people. It is indicative of the virtually unchallenged position which the Prince occupies in the country.

Cambodia has been a monarchy for centuries. As a member of the royal family and a hereditary leader, the Prince commands the loyalty of the rural traditionalists, and his modern nationalism and emphasis on educational and economic modernization have held the loyalty of the urban intellectuals.

2. Political aspects

Through the organ of the single, national organization--the People's Socialist Community--the Sangkum--the Prince controls the political, economic, military, and religious affairs of his country. His protection of the Buddhist hierarchy has insured the support of this conservative element which, in a country whose state religion is Buddhism, provides a basic stability to the nation.

3. International alinement

Prince Sihanouk considers himself a practical nationalist. Realizing his country's inability to resist Chinese hegemony in the area, he made his peace with North Vietnam and China when it appeared that the Chinese- and North Vietnamese-sponsored Viet Cong would be victorious in South Vietnam. While protesting friendship for the Chinese, he nevertheless resisted attempts by the Cambodian Communists to gain power in his government, and, in 1967, threatened to break relations with China due to agitation by members of the Peoples' Republic of China Embassy staff in Phnom Penh. Once the picture in South Vietnam began to change, Prince Sihanouk moved to the other side of the fence.

4. Summary

The Prince can be expected at this stage to be a realist in accepting the facts of international life, but will react in what may appear to be an ill-conceived manner to criticism of either himself or his country.

The style of the country is the style of Prince Sihanouk, and, if he feels that he and his country are being treated properly--as a sovereign nation with democratic rule--he will cooperate in a charming, gracious manner and his people will follow his lead.

Cambodia's national style, as embodied in the Prince and his unchallenged leadership of the majority of the population, is a definite asset to the resistance to Communist-inspired insurgency if the men and weapons are available. Lacking such tools, successful resistance appears extremely questionable, and, seeing the inevitability of a Communist takeover, the Prince might well make a deal with the Communists for what he will honestly believe to be nationalistic reasons—the good of his country and his people.

Annex E (Discussion: National Power) to Strategic Appraisal--Cambodia

1. Discussion question

Does a need exist in Cambodia for the provision of US military and economic assistance?

2. General

The discussion of this question is based on an analysis of Cambodia's assets, material and spiritual, and their applications and adaptability in light of the threat now facing the country, with the object of determining whether Cambodia, without US economic and military aid, can defeat Communist-inspired insurgency and regain control of the countryside. Full development of the considerations of the elements of national power will be found in appendixes to this annex.

The following paragraphs are a condensation of the detailed material concerned with Cambodia's national power in appendixes 1–5 to this annex.

3. Probable favorable effects

   a. Physical Environment. The favorable effects of the physical environment of Cambodia cover several areas. Cambodia is surrounded by nations whose governments are friendly to it, with only the insurgent movements in Laos and Thailand providing an indirect threat to Cambodia. Cambodia has an abundance of arable land completely capable of feeding her population. Its relatively small size and compact shape would be advantageous to military forces having adequate mobility. The tropical climate is not significant. The settlement pattern in Cambodia is unique; almost 95 percent of the population live within 30 miles of a line drawn from the Tonle Sap through Phnom Penh, and on southeastward along the Mekong River. This heavy concentration in a narrow area facilitates the protection of the people and their crops from insurgent exploitation.

   b. Economic System. The favorable effects of the economic system in Cambodia are limited in scope. Its greatest asset is its agricultural productivity, which is heightened by low population pressures. Cambodia is self-sufficient in feeding her population. Despite poor agricultural techniques, Cambodia is normally a major exporter of rice. Second in importance to rice are its large rubber plantations which provide a major source of foreign exchange. Industries that would assist Cambodia in strengthening her military potential are a growing textile industry, a petroleum refinery, a cement plant, and a tractor plant. The major port in Cambodia is Sihanoukville with a capacity of 755,000 tons a year. This port is connected to Phnom Penh by a paved road and a new rail line. Domestic trade is adequate, but is weakened to some extent by the large number of potentially subversive, alien Chinese who control this economic sector. Foreign trade depends almost entirely on rice and natural rubber, as previously noted. The government has instituted policies to improve the economic structure, primarily in controlling the annual trade balance by restricting imports. Cambodia's greatest economic strength lies in her potential for a strong agricultural economy, with supporting light industries.

   c. Political Structure. The governmental structure in Cambodia is a constitutional monarchy with a written constitution. Political power, at least theoretically, is divided among the executive, legislative, and judicial branches. The rights of her citizens are adequately protected. While her influential Chief of State sometimes abrogates this political structure, he has done this to achieve those ends he believes are most desirable for Cambodia and her people. There are also some other weaknesses discussed under the unfavorable effects of the political system. On balance though, when other factors, such as the predominantly agricultural population, low educational level, and traditional acceptance of the King (Chief of State), are considered, the system works quite well for Cambodia under her paternalistic leader. This system has led to political stability. On the surface, this sole reliance on a single leader, Prince Sihanouk, may appear as a disadvantage. However, the Khmer people traditionally accept the leadership of any designated head of the royal family. There is no indication of a lack of core leaders who could take over when Sihanouk dies.

d. Sociology. Cambodia is one of the few nations in Southest Asia which has no population pressures. The needs of the people, particularly those in the rural areas, are unsophisticated and, with few exceptions, are being satisfied. The educational system has been improving since independence. Health standards are rising, but are still inadequate by Western standards. There is a strong national identity and an intense loyalty to the nation which are not only due to her current leader, but extend back to her historical heritage. The same loyalty should relate to any other leader who is a descendant of the royal family. There is no rural landlord class.

e. Military Strength. The regular armed forces of Cambodia consist of around 38,000 personnel. In addition to these regular forces, the armed police and Territorial Defense Force number about 60,000 men. Another source of manpower is a paramilitary youth organization. These supplemental manpower resources could provide sufficient personnel to counter the threat. The regular forces consist primarily of army units of 30 infantry battalions, one armored reconnaissance regiment (battalion), and two parachute battalions. A limited small-craft navy exists, as well as a somewhat larger air force. Cambodia's Chief of State firmly controls the regular forces. The regular forces are responsible directly to him, rather than to the Ministry of National Defense within the Council of Ministers. This strict control could reduce the possibility of a coup by the Council of Ministers. The few Cambodian soldiers trained by the French are good. The Cambodian soldier is brave and loyal, when properly led.

4. Probable unfavorable effects

a. Physical environment. The physical environment also has some unfavorable features. The location of Cambodia, with respect to her remoteness from the United States, would create supply and resupply problems should US military aid be provided. The climate during the wet season will cause equipment to mildew and rust unless protected and maintained properly. Annual flooding of the Tonle Sap Lake increases its size from 1,000 to 4,000 square miles. Considering this flood period and the time required for drainoff, vehicular trafficability in this area and the Mekong Valley is most difficult during three-fourths of the year. These conditions and the accompanying monsoon rains will slow down, but need not stop, aggressive pursuit of the insurgents. The large, forested area and the rough mountains will favor insurgent tactics and require a high degree of training and motivation of government troops.

b. Economic System. Cambodia's economy is weak. Electrical energy resources are low and those that do exist are limited primarily to Phnom

Penh. Known deposits of industrial raw materials are negligible. The labor force is primarily agricultural. Few technically trained personnel are available. It is estimated that there are only 6,000 trained industrial workers, in the Western sense. In addition, many of these are from the Chinese and Vietnamese minorities. Lack of trained personnel hampers the operation and maintenance of the more sophisticated types of military equipment. The financial system in Cambodia is weak. Prices are rising faster than productivity. Each year there have been excessive deficits. The monetary system is, to a large extent, controlled by the Chinese, an unreliable group at this time. Government management of the small industrial sector is inefficient. The telecommunication system is, at best, rudimentary and is concentrated in Phnom Penh. Only one television station and several radio transmitters exist, and they, too, are concentrated in Phnom Penh. The transportation system is underdeveloped. Water transportation is a primary means, particularly in the Tonle Sap/lower Mekong region. Rail lines are limited. In addition to the one connecting Sihanoukville to Phnom Penh, only one other older line exists, connecting Phnom Penh to the Thai rail system. The road net consists of over 5,000 miles with slightly over 2,000 miles being all-weather, much of which is gravel surfaced.

The need for manufactured products keeps rising. This trend could effect a return to an unfavorable trade balance, particularly if rice exports decline. Cambodia's economy received a severe setback with the discontinuance of US military and economic aid.

c. Political Structure. In addition to Prince Sihanouk's abrogating the constitutional process, only one political party exists, the Sangkum (People's Socialist Community). However, varying factions exist within the party which provide grounds for discussion of issues. Cambodia's centralized political organization allows little autonomy at the provincial and district level. Local government officials are appointed and the provincial-elected legislative bodies were dissolved in 1959. The political system in Cambodia is unquestionably authoritarian in character, and some criticism may result from providing US aid to this type of political system.

d. Sociology. As is typical of the population of most emerging nations, there is a large proportion of young people. Some frustration exists with the slowly rising standard of living occasioned by faulty governmental economic policies. The acceptance of inequities, a product of the Buddhist social system, tends to inhibit growth. The Buddhist religion also instills in its followers a nonaggressive attitude, with its attendant effect of suppressing initiative. While the educational system is improving, it is oriented more toward a classical, rather than a technical system; the latter is essential for industrial development. The literacy rate is low. The minority groups control key sections of commerce and industry. An increasing number of educated rural young people have migrated to Phnom Penh. This group is a potential source of unrest due to unemployment and/or the rising cost of living.

e. Military Strength. At the present time, the regular army forces are committed to fighting the insurgents. Additional personnel to expand the army are available from such sources as the Provincial Guard, the Comchivapol, and the paramilitary youth organization. These personnel are generally poorly trained. Intensive training and military equipment will be required to defeat the insurgent forces. Other than the few French-trained soldiers, training of the Cambodian Army, particularly unit training, is poor. Cambodian soldiers come from rural families, are young, and poorly educated. There are few trained leaders. Logistics are adequate for peacetime operations, but are believed inadequate for expanded military operations without an increase in properly trained personnel. Maintenance is marginal, at best.

TWELVE

THE MILITARY PROFESSION AND POLITICAL POWER: RESOURCES AND THEIR MOBILIZATION

Contemporary social science offers a variety of definitions of power. In many or most of these definitions, the central feature is the overcoming of resistance. For example, Dahrendorf, following Weber, defines power as "the probability that one actor within a social relationship will be in a position to carry out his own will despite resistance, regardless of the basis on which this probability rests".[1] Etzioni, after noting that the realization of a societal goal requires the introduction of a change in social relations and that attempts to do so usually encounter some resistance, sees power as "a capacity to overcome part or all of the resistance, to introduce changes in the face of opposition".[2] In Blalock's formulation, power refers to "the actual overcoming of resistance in a standard period of time".[3] Finally, Dahl means by power the capacity of an actor A to get another actor, B to do something that B otherwise would not do.[4] Here, the overcoming of resistance is implied rather than explicit.

The processes of power are most clearly visible in actual decision-making (although, as pointed out in the previous chapter, these processes do not represent an exhaustive set of expressions of social power). This is a central concern to Dahl, who emphasizes the necessity of studying "key decisions" in order to arrive at reliable inferences regarding the actual exertion of political power.[5] And Lasswell and Kaplan state concisely that "power is participation in the making of decisions; G has power over H with respect to the values K if G participates in the making of decisions affecting the K-policies of H."[6]

I shall define *military political power* as the overcoming of resistance in the making of decisions concerning objectives (see the previous chapter) that have (perceived or actual) consequences for the military establishment. Following Blalock, power can further be defined as a multiplicative function of resources and mobilization.[7] As implied by this formulation, both factors have to be present for power to exist. For example, weapons alone do not make an armed intervention; they have to be mobilized, i.e., utilized by some actor who is motivated to use them, who expects that he will have at least some success, and who has some concrete objective to achieve through his actions. Mobilization is motivated behavior. It will be feasible to define mobilization as a multiplicative function of values, expectancy, and objectives.[8]

[1] Ralf Dahrendorf, *Class and Class Conflict in an Industrial Society*, London, 1959: Routledge and Kegan Paul, p. 166

[2] Amitai Etzioni, *The Active Society*, New York, 1968: Free Press, p. 314

[3] Hubert M. Blalock, *Toward a Theory of Minority-Group Relations*, New York, 1967: Wiley, p. 110

[4] Robert A. Dahl, "The Concept of Power", *Behavioral Science*, 2, July, 1957, pp. 201–215

[5] Robert A. Dahl, "A Critique of the Ruling-Elite Model", *American Political Science Review*, vol. 52, June, 1958, pp. 468–69

[6] Harold D. Lasswell and Abraham Kaplan, *Power and Society*, New Haven, 1950: Yale University Press, p. 75

[7] Blalock, *op.cit.*, p. 110

[8] Cf. *ibid.*, pp. 126–127, and below, footnote 20.

1. Resources

"Resources" are those properties of the military group that provide the potential to exert power, although are not sufficient conditions for power exertion. Two aspects of resources seem especially noteworthy for the analysis of the military's ability to influence political decisionmaking, namely, (1) structural and (2) quantitative aspects. Some examples of both will be given below, although without pretensions to completeness. The two kinds are sufficiently distinct from each other to warrant separate treatment, although it will be found that some structural factors cannot readily be considered without their accompanying quantitative component, e.g., 1.1.1 and 1.2.3.

1.1 Structural aspects refer to the relations of the military with the organs of the state, with other politically significant groups, and to the relationships among military sub-groups and cliques. Some important variables are:

1.1.1 Location of the military in relation to executive, legislative, and judicial positions in the state. The more direct access the military have to such positions, including the case when military officers occupy them, the greater the resources. For example, does the military have direct access to the Chief Executive, or only indirectly (e.g., through a defense department or a war ministry)? Are military officers allowed to campaign for positions in local and central legislatures? How widespread are the jurisdictions of military courts? What types of offenses are subject to military trial?

1.1.2 The pattern of interaction between the military establishment and dominant civilian groups, such as business, industrial, and scientific leaders, journalists, educators, labor and management leaders. The closer and the more intimate these connections, the greater the resources for exercising political power.

1.1.3 The existence of para-military forces, and the relationship between them and the military. The closer and more intimate the connections of the military with the police, militia, civil guards, private armed forces, patriotic youth organizations, women's military organizations, etc., the greater the military's resources for the exertion of power. Note however, that to the extent that para-military forces have goals differing from those of the military, they may constitute a threat rather than a positive resource; cf. the relationship between the S.A. and the regular military in Nazi Germany. In 1934, the army agreed to support Hitler for President; the latter "acquiesced in the suppression of Röhm and the S.A., who had dreams of replacing the Reichswehr with a mass, ideologically oriented, people's army".[9]

1.1.4 The relative unity or disunity of the military establishment, i.e., the degree of inter-service and inter-program rivalries, and intra-service factionalism. In general, the greater this degree, the lower the resources (ceteris paribus). (It should be noted, however, that a government decision to favor a particular service, causing inter-service conflict, may increase the probability of direct military involvement in politics. But

[9] Samuel P. Huntington, *The Soldier and the State,* Cambridge, Mass., 1957: The Belknap Press of Harvard University Press, p. 113

this would be due to increased *motivation* on the part of the less privileged services, and hence has to be referred to the mobilization component).

1.2 Quantitative aspects have to do with the concentration of economic funds, manpower, and equipment within the military establishment. Some examples are

1.2.1. The *proportion of the gross national product* spent on the maintenance of the military.

1.2.2 The *military participation ratio* (MPR), or the proportion of militarily used individuals in the total population.[10]

1.2.3 The proportion of military *officers in advisory and executive positions.*

1.2.4 The proportion of industrial *production* devoted to the *manufacturing of military goods* (arms, ammunition, clothing, communications equipment).

1.2.5 The *proportion of the total research-and-development budget* used for military purposes.

1.2.6 The amount of *weapons, transportation,* and *communication equipment* per manpower unit (rifles, guns, tanks, armored cars, cars, buses, trains, ships, aeroplanes, telephones, switchboards, radio, telegram, and teletype machinery, etc).

Given a certain degree of mobilization, the larger these factors, the greater the political power of the military.

Needless to say, the *patterns of utilization* of these resources, i.e., the methods employed to influence decisions, may vary depending on the situation. Military political activity has many facets, and political pressure may be applied through, for example,

(a) *Legislative liaison* (lobbying), i.e., the assignment of personnel for contacts with political, leaders, especially those who have major influence over defense affairs.

The following are some figures related to military lobbying efforts in the US Congress. In 1958, the Department of Air Force Office of Congressional Liaison had 137 persons (55 officers and 82 civilians) "or more than the number of personnel on the congressional staff used in civilian control" (Janowitz, *The Professional Soldier,* p. 358). In 1969, 339 Defense Department employes were assigned to legislative liaison (*Look,* August 26, 1969, p. 18).

(b) *Public statements,* speeches, books, pamphlets, etc., aimed at clarifying the military's arguments with regard to a certain objective. Note, for example, Thomas S. Power, *Design for Survival,* New York, 1965: Coward-McCann, Inc.; Nathan F. Twining, *Neither Liberty Nor Safety,* New York, 1966: Holt, Rinehart, & Winston; Curtis E. LeMay and Dale O. Smith, *America is in Danger,* New York, 1968: Funk and Wagnall.

(c) *Collusion* with groups opposing the government in order to undermine its resistance to military proposals, for example, as in the

[10] Stanislav Andreski, *Military Organization and Society,* London, 1968: Routledge and Kegan Paul. For computations of the MPR for a large number of countries, see Bruce M. Russett et al., *World Handbook of Political and Social Indicators,* New Haven, Mass., 1964: Yale University Press.

demands by the Spanish military that LaCierva should be instituted as Minister of War in 1917. This demand forced Prime Minister Dato to resign.[11]

(d) *Threats of or actual resignation from service* by military leaders. For example, Hindenburg's and Ludendorff's threat to resign if the Emperor did not force Bethman-Hollweg to relinquish his post as Chancellor.[12] Of somewhat less historical significance is the resignation by the Swedish Air Force Chief of Staff, General C. H. Nordenskiöld in 1969, giving as reason for his decision that "there is in Sweden a growing tendency to make optimistic judgments in security policy, ignoring the actual possibilities of armed conflicts".[13]

(e) *Refusal to protect the government from violence*, i.e., the refusal by General Song Yo Chan, Chief of the South Korean General Staff, to defend President Syngman Rhee against student opposition, thereby forcing his resignation (1960).[14]

(f) *Formation of special pressure groups* to support military demands, for example, the Spanish *Juntas de Defensa*, politically active between 1917 and 1923.[15]

(g) Overt *rebellion*, using armed violence (military intervention).[16]

Examples of type (g) are rare in the history of industrialized nations, whether Eastern or Western. At least in part, this may be explained by the fact that they involve *force* rather than power. As Bachrach and Baratz have pointed out, there is an essential difference between power and force in the sense that the exertion of power, aiming at the opponent's compliance with one's wishes, retains a certain freedom of choice for the opponent; whereas in the exertion of force, one's objectives have to be achieved in the face of *non*compliance.[17] If force is used, one drastically reduces his possibilities of utilizing the opponent's further services. Therefore, as long as the military's political objectives are limited and can be met through the cooperation of the existing political leadership, coercive means will usually be avoided in favor of less forceful means of interference. The "actual application of sanctions is an admission of defeat by the would-be wielder of power", as Bachrach and Baratz point out; so it is "to the extent that the prior *threat* of sanctions failed to bring about the desired behavior".[18] In other words, overt, armed rebellion contains considerable risks, not only in the sense that it may be defeated, but also because it reduces cooperation from political and civic leaders and — at least in countries without traditions of military intervention, e.g., the Western industrialized nations as compared to Latin American republics — may call forward strongly unfavorable reactions from the public (for instance, in a general strike). Other ways of exerting power are more readily available; they are also less conspicuous, and therefore less likely to meet resistance.

As a corollary, in less economically developed cultures lacking an institutionalized political system, with few major organizations, a high degree of illiteracy, and without an elaborate system of industrial,

[11] See S.F. Finer, *The Man on Horseback*, London, 1962: Pall Mall Press, ch. 10

[12] See Gordon A. Craig, *The Politics of the Prussian Army 1640–1945*, Oxford, 1955: Clarendon Press, pp. 313–326.

[13] *Dagens Nyheter*, November 2, 1969, p. 3.

[14] Finer, *op.cit.*, ch. 10

[15] *Ibid.*, loc.cit.

[16] There seems to be no point in giving examples here. The literature abounds with case studies of military coups. See e.g. the bibliographies by Moshe Lissak in Morris Janowitz (ed.), *The New Military*, New York, 1964: Russell Sage Foundation, pp. 339–362, and Kurt Lang, *Sociology of the Military*, Inter-University Seminar on Armed Forces and Society, 1969, esp. pp. 57–74.

[17] "Decisions and Non-decisions; An Analytical Framework", *American Political Science Review*, vol. 57, 1963, p. 636.

[18] *Ibid.*, loc.cit

business, and scientific organizations, the alternatives to open rebellion are more scarce. Since there are few other channels for the exertion of power, the military wanting to affect internal conditions in order to achieve greater advantages will tend to use the coup weapon more frequently.

A frequent explanation of military coups in the Third World is that the military constitutes a functional alternative to civil administration. Irving Horowitz states that

The rapidity of the developmental process guarantees the absence of well-organized, popularly controlled political parties. In the absence of such parties, and at times of crisis, the armed forces may be the only group capable of maintaining political order or preparing the ground for further economic breakthroughs. Thus the military tends to perform all sorts of omnibus functions, paramount of which is direct political rule.[19]

[19] *Three Worlds of Development*, New York, 1966: Oxford University Press, p. 268.

The points developed previously suggest that it is not simply the absence of political parties that increases the probability of coups d'etat, but that coups will rather be consequences of the lack of economic and social differentiation in general. Further, the assumption that the military interferes in politics motivated by a desire to maintain political order or to prepare for "economic breakthroughs" is highly questionable. The results of a coup d'etat is often not primarily more order, but simply a different ruling clique, and the successes of the Latin American military governments in achieving economic growth are hardly impressive.

In sum, *if* economic and social differentiation emerges, military rebellion and armed intervention may become more rare; but, if so, it may again be erroneous to draw the conclusion that military *power* is absent just because there are few expressions of military *force*.

It may be wrong to infer that a military coup d'etat in, say, Honduras or Paraguay, represents a greater impact on economic and civic life than, for instance, lobbying efforts of the US military for a new weapons system such as the ABM. Because of the wide networks of economic, scientific, educational and other contacts of the military establishment in industralized nations, the military can exert social and political power in a multitude of ways that are excluded to the military in less developed nations. Therefore, international comparisons of military political power between industrialized and non-industrialized societies is a highly complicated business; and it may be incorrect to claim that the military in a non-industrialized country A has greater power or interferes more than the military in an industrialized country B, on the ground that A has experienced a number of interventions and B has not.

2. Mobilization

Mobilization is motivated behavior. Motivational considerations are important in analyses of political power, since judgments on whether a particular group has power or not basically is dependent on assumptions concerning whether the group *wants* to exert power or not. As we shall see (sec. 3), such assumptions are central in explaining the conflicting arguments of social scientists regarding the political role of the post-World War II American military.

Mobilization may be defined as a multiplicative function of *values, objectives,* and *expectancy*.[20] The two major components to be discussed here are values and objectives, whereas expectancy, i.e., the subjective probability that a desired objective will be achieved, will largely be left uncommented. It should be noted however, that the strength of the expectancy component in military power is inversely related to the degree of civilian control, of which a few remarks will be made later on (ch. 13).

Regarding the concepts of value and objective, the reader may want to refer back to ch. 11, sec. 1. The degree of internalization of professional values will be related to the degree of corporateness and cohesion of the officer corps. Objectives are "characteristics of the real world",[21] i.e., the direct rewards and punishments stemming from concrete situations. Values are more basic to the professional culture and less subject to change than objectives. As discussed in Part II of this book, values are brought about by the homogenizing processes contained in the concept of professionalization$_2$.

One implication of the multiplicative requirement is that mobilization does not occur for values only; in other words, there will be no mobilization unless there is a specific objective to take action *for*. Conversely, objectives alone will not bring forward mobilization unless they are accompanied by values.[22] This is consistent with the point raised in the previous chapter, where objectives were seen as related to and functioning as concrete instances of values.

Since professional values have been extensively discussed in Part II, I can be fairly short here. Two points should be raised, however. First, mobilization of resources will be particularly likely when the objectives of the military are perceived as consistent and intimately connected with values like nationalism, alarmism, and pessimism about the nature of man. Finer's emphasis on the military's identification with the "national interest" exemplified by von Seeckt's idea that the army should be "the purest image of the nation" or Péron's statement that the armed forces are "the synthesis of the nation" is in line with this hypothesis.[23] Second, if was hypothesized in Part II that the more professionalized a certain group, the more "pure" and coherent its cluster of professional values. In other words, the greater the degree of professionalization, the greater will be the tendency of "high" values of each dimension to cluster together. This will increase the size of the value component in mobilization and, with other components constant, this means that the *mobilization of resources will be larger the greater the degree of professionalization*. This implication is contrary to Huntington's argument that professionalization *diminishes* the tendency of the military to participate in politics. (I shall examine Huntington's theory, and its implications for military political neutrality, in the next chapter).

Most of the *objectives* of military political activity lie within the field

[20] This definition is parallel to Blalock's, *op.cit.*, p. 127. I have found it convenient to use slightly different terms, however; thus "value" is substituted for Blalock's "motive", and "objective" for Blalock's "incentive". These changes do not alter the meaning of the definition in any significant way.

[21] *Ibid.*, p. 34.

[22] Expectancy is assumed constant and greater than zero.

[23] Finer, *op.cit.*, ch.4, sec. 2. As Finer points out, military men often strongly stress their responsibility to the nation-state at the expense of loyalty to the existing government. (For some historical examples, see *The Man on Horseback*, pp. 25–26). A clear exposition of this attitude is MacArthur's dictum: I find in existence a new and heretofore unknown and dangerous concept that the members of our armed forces owe primary allegiance or loyalty to those who temporarily exercise the authority and its constitution which they are sworn to defend. (Quoted in Telford Taylor, *Sword and Swastika*, p. 354). Similarly, Johnson points out about the Latin American officers: "they reserve the right to support the constitution while refusing to support politicians" (*The Military and Society in Latin America*, Stanford, 1964: Stanford University Press, p. 252.)

of resource allocation to the armed forces. Larger appropriations mean more weapons and materiel, higher salaries, and better promotion opportunities.

However, because of the growth of what Janowitz has termed "ancillary" military functions — such as military government in occupied countries, military assistance programs, and police functions — and as a consequence of the extension of logistical services, new organizational forms, etc., the number of areas of potential military interest has tended to increase sharply, especially during the present century. As a result, we find a great variety of fields within which the military may — and frequently do — judge it imperative to exert power.

The following systematization of such fields should be regarded as a preliminary overview only. New incentive areas may have to be added, and each one of those indicated below can of course be sub-divided in a number of more specific categories.

First, in the field of *foreign policy* the military often have interests in the outcome of decisions regarding, for example,

alliances with foreign countries (proposing or joining)

foreign loans (giving or accepting)

arms support (giving or accepting)

troop support (giving or accepting)

other strategic decisions, such as defending or not defending certain areas; deploying or not deploying certain weapons, e.g., nuclear bombs and missiles; policies regarding arms control, etc.

Second, with regard to *domestic policy* the military may pursue a multitude of objectives related to, for example, the following fields (the first item is of course selfevident):

planning of, resource allocation to, and education of personnel for the armed forces (e.g., issues pertaining to organization, distribution of resources between the services, recruitment of manpower, curricula of military schools);

the country's economic infrastructure (e.g., the planning of roads, railroads, waterways, air traffic, power plants, harbor installations, localization of industries);

industrial production (e.g., types and amount of production for military purposes, plans for conversion from civilian to war production, domestic production of vital goods in case of isolation);

scientific research (e.g., the proportion of natural science as against humanistic research, output of research with (potential) military applications such as in missile technology and CBR warfare, output of economic, geographical, anthropological, sociological, and political information on foreign countries);

public education (output of technically trained students, output of militarily trained students (for example, through ROTC courses at civilian universities), educational programs with information on the role of the armed forces, etc);

dissemination of information through mass media (e.g., publication of military material in newspapers, radio, television, films, influence over the release of classified material, influence over media material with potential consequences for public morale and "defense spirit");

relations with labor and management organizations (e.g., in order to solicit support for the establishment of defense industries, and to restrict labor-management conflict in strategic industries).

Other things being equal (values, and expectancy) the larger the number of objectives in areas such as these, the larger the total amount of mobilization and, with given resources, the greater the political power of the military. Considerations of civilian control will have to take into account not only the military's influence in the allocation of military resources to the armed forces, but also the extent to which it is able to affect the outcome of decisions in a large number of other subject areas. A "constitutional" analysis of military power is insufficient and may be highly misleading, since the influence of the military in modern society is multi-faceted and dispersed over many areas not foreseen in constitutional rules designed to regulate civil-military relations.

3. The role of motivational assumptions in social science research on the military

As pointed out above, judgments on whether the military has political power or not depends to a considerable extent on the assumptions one makes about the value component in mobilization, i.e., hypotheses concerning whether the military actually desires to exert power or not.

That such assumptions have consequences for the interpretation of events in the civil-military arena can best be illustrated by a comparison between research works on the military. For that purpose, I have chosen publications by two well-known American social scientists, namely, Samuel P Huntington's article "Power, Expertise and the Military Profession"[24] and Morris Janowitz's *The Professional Soldier*. Their common object of study is the political behavior of the American military in the post- World War II period. As we shall see, the authors come out with widely diverging interpretations, in spite of the fact that they to a large extent recognize the same factual data; indeed, I venture that the main differences between them lie not so much in different uses of empirical material, but more in differences regarding their assumptions on military power aspirations.

[24] In Kenneth S. Lynn (ed.),*The Professions in America*,Boston, 1963: Beacon Press. pp. 131–153

To start out with Huntington, he admits that during the early years of the Second World War, many generals and admirals moved into politically powerful positions. Since then, however, there has been an almost continuous decline in the power of the military, depending on five factors:
1) decreased political influence of World War II military commanders;
2) fewer military officers appointed to top political positions under presidents Eisenhower and Kennedy than under Truman;
3) increasing expertise and influence of civilians in the formation of military policy;

4) centralization of authority over military policy in the executive branch;
5) continued divisions among the military (interservice controversy, intraservice divisions, interprogram rivalries).

Despite some allegations of inadequate civilian checks on the military (e.g., former President Eisenhower's famous memento on the military-industrial complex)[25] civilian control, according to Huntington, is effective.

> The image of military dominance is false and dangerous. In actual fact, the power of the military profession 'in the councils of government' has decreased steadily since World War II. It reached its postwar nadir under the vigorous leadership of President Kennedy and Secretary of Defense McNamara at the very time, paradoxically, that concern about the growth of the military was on the upspring.

How is it then that some — like Eisenhower — see a growth in military influence when there is actually a decline? Huntington contends that such interpretations are often based on quantitative rather than institutional criteria,[26] e.g. concentrations on the defense budget rather than on the dispersion and localization of military leaders.

In contrast to Huntington, Janowitz points to an increase in the power of the military after 1945. He attributes it largely to the changing role requirements of the military, brought about by changes in the technological and political environment. Technological innovations, new organizational forms, and military alliances have worked towards a gradual inclusion of political perspectives in the set of professional outlooks. After World War II,

> ... the new international coalition required immediate involvement of the military establishment in many quasimilitary operations, and in direct political administration as well. The military establishment had become a multi-purpose organization in which 'ancillary' functions — military government, military assistance programs, political propaganda, and police functions — assumed great importance.[27]

The power position of the military is seen as a function of both environmental factors and indigenous professional perspectives. Certain conditions have tended to limit civilian control:
1) intensification of inter-service rivalries (since this decreases the applicability of centralized power);
2) congressional supervision of an essentially budgetary kind — the role of Congress as a forum for debating and reviewing national security policy is limited;
3) intensification of military attempts to gain access to the chief executive and the National Security Council.[28]

After this presentation of contrasting viewpoints by Huntington and Janowitz, we may study the role played by different assumptions on power aspirations by comparing two details in the authors' arguments: first, the decreasing role of Congress in military supervision and a correspondingly increasing role of the executive branch; second, divisions among the military in the form of e.g. interservice rivalries. While both writers recognize these developments as being important, they differ

---

[25] "We must guard against the acquisition of unwarranted influence, whether sought or unsought, by the military-industrial complex. The potential for the disastrous rise of misplaced power exists and will persist." "We must never let the weight of this combination endanger our liberties or democratic processes.... Only an alert and knowledgeable citizenry can compel the proper meshing of the huge industrial and military machinery of defense with our peaceful methods and goals, so that security and liberty may prosper together". (January 17, 1961).

[26] Huntington's term "institutional criteria" corresponds closely to what has been called "structural aspects" of resources in sec. 1 above.

[27] *Op. cit.*, pp. 303–304

[28] *Ibid.*, ch. 17

strongly in the role attributed to the military. Huntington points to a "centralization of authority over military policy in the executive branch";[29] Janowitz describes the same process as "an intensified struggle /by the military/ to gain access to the pinnacle – to the chief executive and to the National Security Council".[30] On the topic of interservice conflict, Huntington says:

[29] *Op.cit.*, p. 146

[30] *Op.cit.*, p. 350

Interservice controversy, intraservice divisions, interprogram rivalries all helped to weaken the voice of the military. On few, if any, major issues did the military professionals develop a coherent military viewpoint. Split among themselves, they invited civilian intervention into military affairs.[31]

[31] *Op.cit.*, p. 147

While Janowitz recognizes the same factual premises, his judgment is that the absence of "effective arrangements for realistic unification of the military establishment" has tended to increase rather than decrease "the involvement of the military profession as a pressure group in the domestic political arena".[32] Behind official service positions,

[32] *Op.cit.*, p. 349

... the professional and personal attitudes of ranking officers operate to influence day-to-day decision-making and to supply the matrix through which civilian control must operate.[33]

[33] *Op.cit.*, p. 350

In Huntington's description, the active part in the relationship is the civilian authority, which has taken advantage of weak spots in the military structure to improve its own power position. The military in Huntington's analysis tend to take on the quality of a passive, neutral, and relatively easily governable body of expert-servants.[34] Janowitz on the other hand attributes a significant degree of independent action to the military profession, and in his description the military rather than the civilian authority becomes the active part.

[34] Cf. the quotation in ch. 10, footnote 26, where Huntington refers to the political power "that society thrusts upon" American military leaders.

In the choice between the two theories, Janowitz's is clearly preferable since it does not rule out *a priori* the existence of power aspirations among the military. As will be further demonstrated later, Huntington as a matter of definition excludes the possibility that military professionals can be involved in politics.

Thus, as a conclusion, it is clear that motivation is a crucial concept in the analysis of military political power. Further, it is preferable that social scientists work with models that explicitly include this factor, since assumptions on power motives are likely to enter the theory anyway.

Summary and conclusions

1. We have defined military political power as the overcoming of resistance in the making of decisions that have (or may have) consequences for the military establishment.

2. Military political power is a multiplicative function of (a) resources and, (b) their mobilization, the latter again a multiplicative function of values, objectives, and expectancy. Expectancy is inversely related to the degree of civilian control (see ch. 13); values and objectives are brought about by professional socialization and other mechanisms discussed in Part II.

3. Resources are of two kinds, (a) structural and, (b) quantitative. The former have to do with, for example, the location of the military in relation to executive, legislative, and judicial positions in the state; the latter with, e.g., the amount of men, materiel, and economic investments within the military sector. Some patterns of utilization of these resources were explored, and an explanation of the differences in the frequency of military coups between developed and less developed countries was offered, on the basis of differences in economic and social differentiation.

4. A variety of fields containing potential political objectives for the military profession were enumerated. It was concluded that civilian control cannot concern itself only with checking military influence in the allocation of resources to the armed forces, but has to take into account military interests in a multitude of other issue areas as well.

5. Finally, I argued for the explicit inclusion of motivational variables in the analysis of military power, by comparing two analyses of the U.S. military presented by Samuel P Huntington and Morris Janowitz.

THIRTEEN

PROFESSIONALIZATION, POWER, AND CIVILIAN CONTROL: SUMMARY AND CONCLUSIONS

This book has dealt with two main themes: first, military professionalization, second, military power and influence. The concept of professionalization was subdivided into the theoretically separate, but empirically interdependent[1] processes of professionalization$_1$, i.e., the emergence and historical transformations of the military establishment, and professionalization$_2$, i.e., the homogenization of outlooks and behavior.

[1] See Diagram I.

The impact of the military on social life was analyzed with reference to the concepts of normative influence and political power, the first representing the military's ability to affect the dissemination and institutionalization of values such as nationalism and alarmism, and the second its ability to overcome resistance in the process of actual decisionmaking.

It is now time to integrate these four concepts into a more general discussion, and to work out the implications of professionalization for military power and influence. This will be the subject of the first part of this chapter.

Others have suggested various models in order for civilian society to maximize its control over the military establishment. In the second and final part of the chapter one of these models will be taken up for consideration, namely, Samuel P. Huntington's idea that military power aspirations can be eliminated through ensuring maximum professionalization.

1. Professionalization, normative influence, and political power

1.1 Professionalization$_1$: organization and expertise

The historical process of professionalization of the military involves its tranformation from an ascriptively recruited, usually temporarily employed, and — in relation to contemporary standards — low- or uneducated corps of officers, to an achievement-recruited, permanently hired, and well-trained group of experts. These developments are the results of a number of historical processes, such as the emergence of national states; the creation of mass armies; the accumulation of industrial and technological resources through standardization, automation and machine production and inventions such as means of steam transportation, the gasoline engine, the aeroplane, and electronic means of communication; and new organizational forms and managerial innovations, e.g., the division principle, central staffs, and total warfare organizations to integrate military and civilian defense.

The effects of these forces on the military are most comprehensively summarized in the two concepts of organization and expertise. The formal organization arose as a response to the functional differentiation and subdivision of military units forced by new means of warfare. The traditional categories of infantry and cavalry have been superseded by a complex differentiation of various types of infantry and by the replacement of horses with tanks and armored cars. Artillery, engineering, and various supplys units have been introduced, as well as a multitude of staff and command functions.

Today, the specialization of the military establishment is impressive. For example, according to one estimate, American army officers are assigned to about 400 occupational specialties.[2] Only a minority of the personnel in modern armed forces have tasks that directly engage them in combat activities on the battlefield. As Janowitz has shown, the proportion of "civilian-type" occupations (technical, scientific, maintenance, etc) among enlisted personnel in the U.S. Army increased from about 7 per cent at the time of the Civil War to roughly 70 per cent in the post-Korean War forces.[3]

Education developed as a functional corollary to this increasing complexity. Military academies, often with a strong technical orientation, were established in Europe and the United States around 1800, and in Latin America (under French and German guidance) some one hundred years later. The contemporary curricula of military academies and colleges reflect the variety of professional tasks, and instruction is given on such topics as global strategy, logistics, international affairs, government and politics, economics, decisionmaking and problem solving, and public speaking and presentation techniques.

Organization and expertise are major factors underlying both the military's normative influence and its political power.

Normative influence. As the military became consolidated as a symbol of national unification and supremacy, and as it developed into an instrument of the state rather than of individual power-holders (such as feudal chiefs), the ideology of the officer corps gradually changed. Soldiers' loyalty to the military system was insured not on the basis of pay, as in the older, internationally recruited mercenary forces, but on the basis of attachment to the common national cause, and with the aid of new laws of general conscription. The transition from the mercenary

[2] Samuel P. Huntington, "Power, Expertise, and the Military Profession", in Kenneth S. Lynn, (ed.), *The Professions in America*, Boston, 1967; Beacon Press, p. 132.

[3] Morris Janowitz, *The Professional Soldier*, Glencoe, Illinois, 1960: Free Press, p. 65. See also the comparison between the U.S. military and civilian labor forces in Harold Wool, *The Military Specialist*, Baltimore, 1965: The Johns Hopkins Press, p. 52

to the national stage meant the appearance of a new set of nationalistic doctrines. But it also was accompanied by a concentration of scientific, industrial, and technological resources to the military establishment, and hence contributed to connecting the military establishment with a variety of civilian and economic groups ready to invest their resources in the production of military materiel and equipment.

Professionalization is relevant to normative influence in several ways. First, it established the military as an important pressure group for the dissemination of values such as nationalism, alarmism, and distrust vis-a-vis other nations. Secondly, technological development and the increasing complexity of military planning has offered a rationale for the *continuous* existence of war preparations and weapons development, i.e., organization on a permanent basis rather than ad hoc efforts in times of crisis. This, in turn, exerts a pressure on the military to gain public support and legitimation, expecially during peacetime when military efforts often seem unnecessary to the population at large. Third, the military has become proficient in the utilization of public relations and propaganda, obviously a valuable asset in the spreading of values to wider circles.

Without a permanent formal organization, there might well be an officer corps, but there would be no "military establishment"; the very concept of establishment implies continuity and organization. And as any establishment, the military devotes part of its activity to self-preservation and for providing its own raisons d'etre.

Political power. Professionalization$_1$ has been instrumental in increasing the military's *resources* in various ways. First, with regard to the quantitative aspects, new types of weapons of increasing complexity have been introduced, and their destructive capability has multiplied. As we have seen, one of the decisive stages in the large-scale provision of arms was the solution of the standardization problem and the taking up of serial production around 1800. There is no need to repeat the further stages in the development of weapons again, as this was covered in ch. 1. It is a commonplace, of course, that modern armies do possess a potential of destruction which is far higher than that of earlier periods, even short of the employment of nuclear weapons.

It should be observed that the emergence of the military technical expert introduced an important innovative force in the field of weapons production. Military professionalization$_1$ was associated with the rise of specialists in ballistics, engineering, logistics, etc. As experts, the officers were not content simply with using the arms and equipment created by earlier generations, but they also devoted considerable energy and skill to invent new means of warfare. Prieur, who contributed decisively to the development of mass-production of arms in France through his solution of the standardization problem, was a military engineer. Other French officers, for example, Delvigne, Thouvenin, and Minié, during the first half of the nineteenth century worked on the development of projectiles for guns and rifles; Delvigne established the principle that a rifle bullet should be elongated and ogival rather than spherical.[4] In artillery tactics, the work of du Teil in France was significant for the success of the Napoleonic armies. At West Point, engineers and technicians received education in various specialties useful for weapons production, for example, metallurgy. In England, young officers took active part in the development of the tank. Thus, the rise of the military technical expert was not only a *response* to scientific and technological developments outside the military sphere: it also came to act as a *stimulus* for further accumulation of resources within the military system.

[4] Jacques Boudet et al., (eds.), *Arméernas Världshistoria,* part III, Stockholm, 1968: AB Svensk Litteratur, table 8. Walter Millis, *Arms and Men,* New York, 1956: Mentor Books, p. 82.

Obviously, for a military committed to open intervention, the rapid expansion and modernization of weapons and equipment involves a direct power potential. However, for the political role of the military in industrialized, economically highly developed countries these direct effects are less significant that the indirect ones gained through the extended contacts with, for example, the aerospace, ship-building, and electronics industries.

Thus, professionalization$_1$ involves the gradual development of the structural basis of power resources as well. Besides military-industrial cooperation, there are two other trends that deserve emphasis.

First, as has been noted by several writers, the *transferability of skills* between the military and civilian spheres has increased (at least seen in the perspective of a century), making abilities acquired in military training adaptable to civilian jobs. Engineers, machine maintenance specialists, health service experts, logistic and personnel technicians have relatively good prospects of finding equivalent civilian employment.

More significant, however, is the fact that the increasing similarity between the military and civilian occupational structure makes the military capable of administering a number of societal areas such as communications, transportation, health services, and education. In the event of a military coup d'etat, therefore, the contemporary military does not only command greater coercive resources to overcome civilian resistance, but also resources that enables it to stay in power longer than has previously been the case. In my opinion, Miliband's contention that the general strike is an effective weapon against military intervention overlooks the pervasive character of the *totality* of military functions that parallel civilian ones, and wich may allow the military to substitute their own administration for the civilian one longer than workers may be able to continue their strike. The failure of the Kapp *Putsch* in Germany (March, 1920) is hardly a significant case of negative evidence to this hypothesis, since it was an almost extreme case of confusion and deficient planning, not likely to be repeated by any modern, well-organized officer corps.[5]

[5] Ralph Miliband, *The State in Capitalist Society* (Swedish ed., *Statsmakten i det kapitalistiska samhället,* Stockholm, 1970: Tema, pp. 142–149). On the Kapp Putsch, see e.g., Gordon A. Craig, *The Politics of the Prussian Army 1640–1945,* Oxford, 1955: Clarendon Press, pp. 375–380.

Another by-product of the transferability of skills, significant for the military's political power resources, is the considerable exchange of personnel between the military establishment and civilian posts in industry and government. This issue, raised with such force by C. Wright Mills, continues to attract the attention of social scientists as well as politicians. For example, two investigations have been made in 1959 and 1969 concerning the number of ranking officers (colonels, naval captains, and above) employed by the top 100 defense contracting firms in the United States. In 1959, that number was 721; in 1969, 2 072. According to the 1959 survey, all of the major aircraft and missile industries had retired admirals and generals on key administrative posts; ten of the major companies were reported to employ about half, or 372, of the officers included in the investigation.[6] There is no need for conspiracy theories to suggest that such an "interlocking directorate" (Janowitz) will contribute to the societal power of the military.

[6] Reports by Senators Paul Douglas (1959) and William Proxmire (1969). See Janowitz, *op cit.*, p. 376, and Adam Yarmolinsky, "The Problem of Momentum", in Abram Chayes and Jerome B. Wiesner (eds.), *ABM – An Evalaution of the Decision to Deploy an Antiballistic Missile System,* New York, 1969, Signet Books, p. 148.

Second, the military has achieved *extended political functions.* Although military leaders have acted through history as advisors to kings and princes on matters of war and strategy, it is mainly during the present century the officer corps as a whole – at least in the major world powers – is *professionally* trained and educated to make political appraisals. There seems to be at least three reasons for this development. First, the growing complexity of modern warfare, and the vast political implications of the employment of nuclear weapons. Second, the character of total war, which requires the organization of society as whole for defense purposes and which tends to expand military jurisdiction into a large number of new areas, such as the location of industries, roads, power plants, and the protection of public utilities (water, electricity), transport, and communications. Third, especially in the post-World War II period the military has assumed a number of "ancillary" functions as, for example, military administration of occupied countries, police tasks, counter-guerilla warfare, and training of foreign troops in "civic action". The overall effect of these tendencies is to further erode the distinction between "strategic" and "political" issues; and to the extent that the military succeeds in gaining influence over strategy, claiming this to be "strictly military", it will also continue to gain increased foothold in the political sphere.

1.2 Professionalization$_2$: corporateness and indoctrination

As a short characteristic of the difference between p_1 and p_2 one could say that professionalization$_1$ refers to processes which have led to the establishment of the military as a *pressure* group, whereas professionalization$_2$ turns it into a pressure *group*. Professionalization$_1$ has brought about expertise and the power of organization; professionalization$_2$ involves those characteristics which make that expertise and organization corporate. While p_1 refers to the long-term transformations of the profession, p_2 concerns the mechanisms that create professionals. The ultimate result of professionalization$_2$ is the pure, crystallized, and coherent set of values called the "military mind" (see ch. 4).

Normative influence. As previously pointed out (ch. 11), corporateness and the common recognition of the basic values of the profession are essential elements for its achieving normative influence. The propagation of values can be effective only to the extent that the profession succeeds in minimizing internal dissensus and conflicts over its major goals. To achieve this, the profession exerts control over the selection, socialization, and promotion of members. Professionalization$_2$, then, is a summary concept for mechanisms that ensure smooth succession of leadership, the loyalty of the members to professional values, and which check those opposing forces in the outside world that impinge on the profession and constitute potential threats to its value base.

Political power. While professionalization$_1$ affects the resource component of political power, professionalization$_2$ primarily influences the motivational elements, as contained in the concept of mobilization. Formal and informal socialization processes, such as class education, colleague interaction, reading of journals and memoranda, and attendance at conferences, transmit to the military man those values and objectives he is expected to pursue in this professional activity. Directly or indirectly, they also outline to him the limits of non-compliance, i.e., indicate what degree of tolerance he can expect for attempts at questioning the pattern of basic values. The sanctions meted out for noncompliance are many and various, the most common being direct reproaches from superior officers, and promotions that fail to appear.

As already pointed out in the previous chapter, professional objectives most typically have to do with budget allocations to the armed forces. One of the most serious breaches of corporate ethics would be to publicly express doubts about the profession's appropriations demands; hence, it is extremely rare to find military men demonstrating such objections. In fact, one of the safest generalizations that can be made about professional politics is that the members will almost unanimously support requests for greater resources. A good deal of members' working hours is spent on data collection, research, and policy evaluations in order to support such claims as effectively as possible.

1.3 Summary

The following relationships hold between the four concepts of professionalization$_1$, professionalization$_2$, normative influence, and political power. (see table 11).

A. Professionalization$_1$ is functional to normative influence in the sense that it has established the military profession as a social institution working for the propagation of nationalistic and alarmistic values, and has provided it with the skills and resources required for such tasks.

B. Professionalization$_1$ has implications for political power through the accumulation of resources, both of the quantitative/material kind (note especially the innovative function of professionalization) and the structural variety (transferability of skills, political role-expansion).

C. Professionalization$_2$ contributes to normative influence through the creation of corporate cohesion and solidarity, which are essential requirements for effective external dissemination of values.

D. Finally, professionalization$_2$ has effects on political power through the indoctrination of values and objectives, i.e., two of the three main components of mobilization.

TABLE 11

Implications of professionalization$_1$ and professionalization$_2$ for normative influence and political power

	Implications *for* normative influence	Implications *for* political power
Implications *of* professionalization$_1$	Establishment of the military as pressure group for nationalistic and alarmistic values A	Resources B (a) quantitative (note innovations) (b) structural (transferability of skills, political role-expansion,
Implications *of* professionalization$_2$	Generation of corporateness and reduction of dissensus C	Indoctrination of specific values and objectives D

Let us finally turn to some of the implications of military professionalization for civilian control. In the light of the previous discussion it is clear that civilian control can *not* rely on the maximization of professional characteristics. As we shall see, Huntington's contrary thesis has to be rejected on two grounds: first, because of its inadequate realization of the full effects of military professionalization; second, because of its inconsistencies and formal deficiencies.

2. Huntington: Control through maximum professionalization

Huntington's thesis is developed on the basis of two contrasting concepts of civilian control: "subjective" and "objective". "Subjective" control involves the maximization of the power of one or more civilian groups (such as a governmental institution or a political party) over the military, thereby reducing its autonomy.

The simplest way of minimizing military power would appear to be the maximizing of the power of civilian groups in relation to the military. ... /T/he maximizing of civilian power always means the maximizing of the power of some particular group or groups.... The general concept of civilian control is identified with the specific interests of one or more civilian groups. Consequently, subjective civilian control involves the power relations among civilian groups. It is advanced by one civilian group as a means to enhance its power at the expense of other civilian groups.... In its various historical manifestations, subjective civilian control has been identified with the maximization of power of particular governmental institutions, particular social classes, and particular constitutional forms.[7]

On the other hand, "objective" control has the aim of setting the military free from pressures and counterpressures that may arise from the struggle between civilian political groups. This can be done, because professionalism makes the military uninterested in other matters than those that are strictly professional.

Civilian control in the objective sense is the maximizing of military professionalism. More precisely, it is that distribution of political power between military and civilian groups which is most conducive to the emergence of professional attitudes and behavior among the members of the officer corps. Objective civilian control is thus directly opposed to subjective civilian control. Subjective civilian control achieves its end by civilianizing the military, making them the mirror of the state. Objective civilian control achieves its end by militarizing the military, making them the tool of the state.[8]

Objective control means the political neutralization of the military. "The antithesis of objective civilian control is military participation in politics"; conversely, according to Huntington, subjective control presupposes this involvement by the military.

In short, subjective control implies the effort of civilian groups to restrict military autonomy, while objective control is aimed at widening it. In the first case, civilian control is ensured by the political pressure from civilian groups; in the second, control is guaranteed by the military itself, provided it is properly professionalized.

[7] Samuel P. Huntington, *The Soldier and the State*, Cambridge, Mass.: Harvard University Press, 1957, pp. 80-81.

[8] *Ibid.*, p. 83

2.1 Discussion

Huntington builds his arguments for "objective" control partly on the basis of a derogatory definition of the alternative concept.[9] This, in fact reflects a basic disbelief of his in those forms of control that stem from public concern over the political role played by a large military establishment. "Subjective control" takes on overtones of complaints from groups which lack legitimate reasons for their concerns about the military's political power; the very term "civilian control" may become a "slogan" which is "utilized by groups which lack power over the military forces in struggles with other civilian groups *which have such power*"[10] (italics mine).

Theoretically speaking, this formulation is highly unsatisfactory, since it postulates what Huntington's argument is supposed to prove, namely, that there are groups or institutions that can command sufficient power

[9] One may reflect over the popular connotations of Huntington's concepts. "Objective" implies something universally recognized, intersubjectively valid, stable, and secure, while "subjective" bears the image of something personal, individualistic, idiosyncratic, varying, and unstable. Since few would prefer civilian control having the latter characteristics,

over the military. Presumably, the reason why civilians express concern over of political control is their suspicion that governmental organs are not effective in checking the military's interference in political life. Obviously, if their suspicions are correct, their concerns can hardly be called "slogans".

My second objection has to do with Huntington's use of the term "professionalism". Essentially, a "professional" officer corps is one which exhibits expertise, responsibility, and corporateness.[11] "Professionalism", however, to Huntington also involves political neutrality; as a result, "professionalism" and "objective control" are inseparable as theoretical concepts. The immediate consequence of this is to rule out the empirical possibility of establishing the relationship between the *degree* of professionalism and the *degree* of political neutrality. Huntington's thesis becomes, in Carl Hempel's words, "a covert definitional truth"[12]. In other words, professional officers never intervene; because if they do, they are not true professionals.

As could be expected, Huntington has considerable difficulty to reconcile his historical accounts of civil-military relations with his conceptual scheme. Consider, for example, the following discussion dealing with Nazi Germany.

Huntington points out that the Wehrmacht of the Third Reich had a highly professionalized officer corps. True, there were attempts by the German leadership to penetrate and divide the military by lowering the level of the General Staff, by the creation of the SS as a second army, by the establishment of the Luftwaffe under Göring outside the normal chain of military command, and by the purging of high-ranking officers like Blomberg and von Fritsch.[13]

A great number of top level officers remained, however, filling the posts of those who had been purged. What happened to them?

The reactions of the military to the Nazi penetration split them into three groups. One clique succumbed to Nazi temptations, *abandoned the professional outlook,* adopted Nazi views and were suitably rewarded by the government. Another group, including Hammerstein-Equord, Canaris, Beck, Adam, Witzleben, and most of the 20th of July conspirators, also assumed political roles actively opposing Hitler and his policies. Since both these groups *abandoned professionalism for politics,* it is appropriate to judge them, not by professional, but by political standards. The former share in the guilt of National Socialism; the latter were usually motivated by the highest humanitarian and Christian ideals. ...The great bulk of the officer corps had no political yearnings. ...[14]

In order to show that professional soldiers always obey — consistent with Huntington's main thesis — the author finds it necessary to define away those who did otherwise by saying that they "abandoned professionalism for politics".

There are also other counterexamples to Huntington's argument: The Algiers coup in 1961 was carried out by some of the most professionalized elements in the French Army. Two Sandhurst-educated officers in Pakistan (Iskander Mirza and Ayub Khan) in 1958 conspired to relieve the politicians from power. The Nigerian Army take-over in January, 1966, also was conducted by officers trained at Sandhurst.[15]

Military obedience cannot be made totally independent of the

the hidden recommendation is to prefer the former concept; which is what Huntington does. Intuitively, it may seem more reasonable to switch the terms, exchanging "subjective" for "objective", and vice versa: one may argue that in the first of the quotations above, the military is the objective of political control, whereas in the second one control is a function of subjective processes resulting from professional education, training, and career experiences. It would have been still better to avoid the terms altogether, and use, for example, "directive control" in the first case and "selfregulative control" in the second.

[10] Huntington, *The Soldier and the State,* p. 81.

[11] *Ibid.,* pp. 8–18. "Responsibility" is a dubious term for the description of professions, for reasons already discussed; see ch. 11, sec. 4.

[12] Carl G. Hempel, "The Logic of Functional Analysis", in May Brodbeck (ed.), *Reader in the Philosophy of the Social Sciences,* New York, 1968: Macmillan, p. 193.

[13] Huntington, *The Soldier and the State,* pp. 113–114.

[14] *Ibid.,* pp. 121–122

[15] See Ronald J. Stupak, "The Military's Ideological Challenge to Civilian Authority in Post-World War II France", *Orbis,* Summer, 1968, pp. 582–604; Alfred Vagts, *A History of Militarism,* London, 1959; Hollis

society's political system: it is always paid to *some* group and *some* political ideology. My third objection refers to the lack of correspondence between Huntington's emphasis of the military's professional nationalism, alarmism, and conservatism and, on the other hand, his contention that professionalism involves political sterility. While I agree with Huntington's analysis of the military's professional ideology, I find it impossible to see how it ties in with the notion of the military as being unconditionally obedient, i.e., an officer corps which "stands ready to carry out the wishes of any civilian groups which secure legitimate authority within the state".[16] Apparently, the military's motivation to serve must have some connection with the ideology and policies of the government. There is a fundamental contradiction between Hungtingon's proposal for an independent military sphere and his statement that "/only an environment which is sympathetically conservative will permit American military leaders to combine the political power which society thrusts upon them with the military professionalism without which society cannot endure."[17] If the military thrives on a conservative environment (and there is little reason to question that hypothesis; see ch. 10), then this seems to signify that "objective" control can hardly be made equivalent to political neutrality; and a conservatively biased military professionalism is a far cry from Huntington's notion of a politically sterile military.

General William T. Sherman once demanded the absolute separation of politics from the military, in much the same fashion as Huntington; Sherman wanted to make the Army into "an animated machine, an instrument in the hands of the Executive of enforcing the law, and maintaining the honor and dignity of the nation".[18] But politics and military issues are often difficult, if not impossible, to separate (see the discussion in ch. 11, sec. 5); and if the military is a machine, it possesses a definite capability of influencing its programmers.

As a general conclusion, civilian control cannot rely on the expansion of military autonomy, but rather has to work for its restriction. To check the military's quest for professional independence is important for increasing the political freedom of choice of civilian governments. It should be clear that few patent solutions for eternal peace could be achieved this way; the military is not the only group interested in the exploitation of conflicts to achieve indigenuous goals.

The aim of civilian control has to be expressed in limited rather than utopian terms: the problem is to achieve a degree of control which will leave civilian governments a fair choice as to what methods — military *or* non-military — should be applied to achieve solutions of international crises and conflicts. And there is no point in applying a hard-core social-darwinistic scheme to international relations: states are often as interested as not in exploring détentes and means for coexistence rather than competition and conflict. But with the present strong dominance of the military establishment, governments often find their options restricted by the corporate and well-organized social and political activity of the military.

---

and Carter, p. 490; and Robin Luckham, *Social Structure and Conflict in the Nigerian Officer Corps*, unp. Master's thesis, University of Chicago, 1968.

[16] Huntington, *The Soldier and the State*, p. 84.

[17] *Ibid.*, p. 464

[18] Charles H. Coates and Roland J. Pellegrin, *Military Sociology*, University Park, Maryland, 1965: Social Science Press, p. 211

Robert F. Kennedy, in his memoir of the Cuban missile crisis, has given an account of President Kennedy's views of the reactions of the military leadership as the crisis culminated. President Kennedy was impressed with the effort and dedicated manner in which the military responded by putting the Army, Navy, and Air Force on constant alert. But he was distressed that military representatives with whom he met gave so little consideration to the implications of the steps they suggested.

> They seemed always to assume that if the Russians and the Cubans would not respond, or, if they did, that a war was in our national interest. One of the Joint Chiefs of Staff once said to me (R.F.K) he believed in a preventive attack against the Soviet Union. On that fateful Sunday morning when the Russians answered they were withdrawing their missiles, is was suggested by one high military adviser that we attack Monday in any case. Another felt that we had in some way been betrayed.
> President Kennedy was disturbed by this inability to look beyond the limited military field. When we talked about this later, he said we had to remember that they were trained to fight and to wage war – that was their life. Perhaps we would feel even more concerned if they were always opposed to using arms or military means – for if they would not be willing, who would be? But this experience pointed out for us all the importance of civilian direction and control and the importance of raising probing questions to military recommendations.[19]

[19] Robert F. Kennedy, *Thirteen Days*, New York, 1969: Signet Books, p. 119.

Although, fortunately, most problems of civilian control arise on less fateful occasions than crises involving threats of immediate nuclear confrontation, the passage brings out one thing very clearly: civilian control hinges on an explicit recognition of the military's professional bias, and on the adequate capacity to respond to that bias.

Such capacity does not only – and not even primarily, although the Cuban crises centainly is an important exception – depend on the wisdom and consideration of individual political leaders. The impact of the military establishment is based on organization and on the expertise of the men who run it; hence civilian control, in order to be effective, has to take the professional characteristics of the military into account, and to devise means to deal with their consequences. In this endeavor, the effectiveness of current methods of control and supervision has to be carefully estimated.

One of the chief means for civilian control in multi-party states is legislative supervision, carried out by organs such as parliamentary and congressional committees, primarily – although not generally – aimed at the scrutinization of various aspects of the military budget. One of the working assumptions of this method is that the members of the legislative body have enough motivation and time to carefully consider the various posts of the budget, relating them – or at least the major ones of them – to the long-term strategic goals of the armed forces.

This assumption does not seem to hold, for at least the following reasons. First, politicians in multi-party states typically find the electorate disinterested in matters of security policy, and are thus tempted to prefer the exploration of other issues for potential voting turnouts. Second, politicians are often reluctant to place themselves under a possible attack from competitors, and from the military, for not being sufficiently "patriotic". Third, and most important, budgetary reviews of billion or multi-billion dollar budgets is an exceedingly complex undertaking. For example, as Janowitz has pointed out with regard to the U.S. Congress, "it is the uniform conclusion of various researches into the congressional review of the military budget that it is an outmoded technique of rather limited consequence. Its effect on the military profession seems to be that of generating hostility and tension rather than effective control and political consent." This view is substantiated by some figures related to the results of budgetary control: "congressional efforts at reducing the military budget in the years since World War II have resulted in a reduction of less than 5 per cent from recommended levels".[20] Janowitz contends that congressional committees have found themselves unequal to the technical problems, and have depended upon information from the military and industry.

It is in the light of such findings one has to see Galbraith's proposal for the forming of a group of civilian experts and scientists, who should work as advisors to the political organs and the Congress, providing an alternative to military expertise. Such a group, if adequately organized, would constitute a very important complement to the more traditional means of civilian control. It could work out alternative solutions to security problems, without applying the alarmist bias of the military. It could act as a permanent organization in order to utilize signs of de-escalation and decreasing tension for further de-escalation; and it could work out solutions for the conversion of the production of war materiel to civilian production.

As was pointed out in ch. 11, the pluralist theory of elites is inadequate in not recognizing that a foremost source of power is the dominance in a certain issue area; hence an important means to achieve the "cancelling out" effect of pluralist theory is to establish competing expertise in fields previously dominated by one group only.

An organization such as the one outlined by Galbraith would have three main functions. First, it would contribute to the institutionalization of values contrary to those of the military establishment (provided it is given adequate resources for publishing its programs and proposals). Second, it would match military expertise through assisting legislative organs with alternative scientific and technological advice. Third, it would check the military's direct exertion of political power through actively participating in the governmental decisionmaking process. Thus, institutionalization is met by counter-institutionalization, expertise by counter-expertise, and organization by counter-organization.

Professionalization, both in the macro and micro sense, has direct implications for the social power of elites, because of its function as a support and rationalization of their claim to independence in their respective issue areas. As van Doorn has stated, "the concept of military professionalism, like every other kind of professionalism, includes a perennial search for autonomy which is in contradiction to tight political control."[21] Hence, if civilian society is to retain dominance over the

[20] Janowitz, *op.cit.*, pp. 354–355

[21] Jacques van Doorn, "Political Change and the Control of the Military", in *ibid.*, (ed.), *Military Profession and Military Regimes,* Mouton, 1969: The Hague, p. 12.

military, rather than the other way around, it cannot afford to be negligent about the effects of professionalization. Terms like "scientific objectivity" and "moral neutrality" are inadequate when describing a profession which is so deeply involved in the political process. Professionalization creates experts, but it also gives them resources, corporate interests, and objectives to pursue; moreover, it does not guarantee that those interests and objectives are consonant with the wants and needs of the public, or even of the state's leadership.

Because of its pervasivensess and its economic and political impact, the military will continue to be one of the main issues of the political debate; and the profession can hardly be expected to let that debate go unnoticed. Like other pressure groups, it will strive to implement values and fulfill objectives that it defines as central to its self-preservation. This is the reality which civilian society has to face and for which appropriate means of control ultimately will have to be provided.

REFERENCES

Abrahamsson, Bengt, *Anpassning och avgångsbenägenhet bland militärt befäl,* Stockholm, 1965: Militärpsykologiska institutet, report no. 37.
— "The Ideology of an Elite", in van Doorn, J.A.A. (ed.), *Armed Forces and Society,* The Hague, 1968: Mouton.
"American Militarism", *Look,* 1969, vol. 33, nos. 16 and 17 (August 12 and 26).
Andreski, Stanislav, "Conservatism and Radicalism of the Military", *European Journal of Sociology,* 1961, no. 1.
— *Military Organisation and Society,* London, 1968: Routledge and Kegan Paul.
Angående möjligheten och önskvärdheten av fred. (Report from Iron Mountain), Stockholm, 1968: Rabén och Sjögren.
Anti-Ballistic Missile: Yes or No? A special report from the Center for the Study of Democratic Institutions, New York, 1968: Hill & Wang.

Bachrach, Peter, and Morton S. Baratz, "Two Faces of Power", *American Political Science Review,* vol. 56, March, 1962, pp. 947–952.
— "Decisions and Nondecisions: An Analytical Framework", *American Political Science Review,* vol. 57 1963, pp. 632–642.
Bacon, Reginald, and Francis E. McMurtrie, *Modern Naval Strategy,* New York, 1941, Chemical Publishing Company.
Baldwin, Hanson W., "When the Big Guns Speak", in Markel, Lester (ed.), *Public Opinion and Foreign Policy,* New York, 1949: Harper & Bros.
Barber, Bernard, "Some Problems in the Sociology of the Professions", in Lynn, K.S. (ed.), *The Professions in America,* Boston, 1965: Beacon Press.
Barnes, John W., "Logistics", *Collier's Encyclopedia,* New York, 1962.
Barnett, Corelli, "The Education of Military Elites", *Journal of Contemporary,* 1967, vol. II, no. 3, pp. 15–35.
Beaufre, André, *Modern strategi för krig och fred,* Stockholm, 1966: Prisma.
Befälsordningen vid infanteriet, Stockholm, 1953:SOU 1953:28.

Bendix, Reinhard, and Seymour M. Lipset, "Political Sociology", *Current Sociology*, vol. VI, 2, pp. 79–98.

Bengtsson, Eva-Stina, "Some Political Perspectives of Academic Reserve Officers", *Journal of Peace Research*, no. 3, 1968, pp. 293–305.

Benoit, Emile, and Harold Lubell, "The World Burden of National Defense", in Benoit, Emile (ed.), *Disarmament and World Economic Interdependence*, Olso, 1967: Universitetsforlaget.

Blalock, Hubert M., *Toward a Theory of Minority Group Relations*, New York, 1967: Wiley.

Blau, Peter M., and O. Dudley Duncan, *The American Occupational Structure*, New York, 1967: Wiley.

Boalt, Gunnar, "Militärt och civilt. Några sociologiska synpunkter", *Social Årsbok, 1950–51*, Stockholm 1951: KF.

Borup-Nielsen, Steen, *Reserveofficersundersøgelsen 1964*, Copenhagen, 1966: Militærpsykologisk Tjeneste (mimeo.).

Bottomore, T.B., *Elites and Society*, New York, 1964: Basic Books.

Boudet, Jacques, et al. (eds.), *Arméernas världshistoria*, I–IV, Stockholm, 1967–69: AB Svensk Litteratur.

Bradley, Omar N., "Should We Fear the Military?" *Look*, vol. 16, March 11, 1952, p. 35.

Busquets Bragulat, Julio, *El Militar de Carrera en España*, Barcelona, 1967: Ediciones Ariel.

Campbell, D.T., and T.H. McCormack, "Military Experience and Attitudes Toward Authority", *American Journal of Sociology*, vol. 62, 157, pp. 482–490.

Canton, Dario, "Military Interventions in Argentina: 1900–1966", in van Doorn, J.A.A. (ed.), *Military Profession and Military Regimes*, The Hague, 1969: Mouton.

Carlsson, Gösta, *Social Mobility and Class Structure*, Lund, 1958, Gleerups.

Carr-Saunders, A.M., and P.A. Wilson, *The Professions*, Oxford, 1933: Clarendon Press.

– "The Emergence of Professions", in Nosow, S., and W.H. Form (eds.), *Man, Work, and Society*, New York, 1962: Basic Books.

Chayes, Abram, and Jerome B. Wiesner (eds.), *ABM – An Evaluation of the Decision to Deploy an Anti-Ballistic Missile System*, New York, 1969: Signet Books.

Coates, Charles H., and Roland J. Pellegrin, *Military Sociology*, Univerisity Park, Md., 1965: The Social Science Press.

Cook, Fred J., *The Warfare State*, New York, 1962: Macmillan.

Craig, Gordon A., *The Politics of the Prussian Army, 1640–1945*, Oxford, 1955: Clarendon Press.

Cvrcek, J., "Social Changes in the Officers' Corps of the Czechoslovak People's Army", in van Doorn, J.A.A. (ed.), *Military Profession and Military Regimes*, The Hague, 1969: Mouton.

Dahl, Robert A., "The Concept of Power", *Behavioral Science,* 2, July, 1957, pp. 201–215.
- "A Critique of the Ruling Elite Model", *American Political Science Review,* vol. 52. June, 1958.
- *Who Governs?* New Haven, 1961: Yale University Press.

Dahrendorf, Ralf, *Class and Class Conflict in an Industrial Society,* London, 1959: Routledge and Kegan Paul.

Demeter, Karl, *Das deutsche Heer und seine Offiziere,* Berlin, 1930.

van Doorn, J.A.A., "The Officer Corps: A Fusion of Profession and Organization", *European Journal of Sociology,* no. 2, 1965.
- (ed.), *Armed Forces and Society,* The Hague, 1968: Mouton.
- (ed.), *Military Profession and Military Regimes,* The Hague, 1969: Mouton.
- "Political Change and the Control of the Military", in *ibid., Military Profession and Military Regimes,* The Hague, 1969: Mouton.

Douglas, William O., "Should We Fear the Military?", *Look,* vol. 16, March 11, 1952, p. 34.

Durkheim, Emile, *Professional Ethics and Civic Morals,* New York, 1958: Free Press.

Edinger, Lewis J., *Politics in Germany,* Boston, 1968: Little, Brown, and Co.

Encel, S., "The Study of Militarism in Australia", in van Doorn, J.A.A. (ed.), *Armed Forces and Society,* The Hague, 1968: Mouton.

Engels, Friedrich, *The Role of Force in History,* London, 1968: Lawrence and Wishart.
- *Anti-Düring,* Stockholm, 1944: Arbetarkultur.

Ericson, Stig H:son, "Att vara svensk officer", *Fred och försvar under 60-talet,* Stockholm, 1961: Esselte.

Eriksson, Robert, *Yrkesval och officersrekrytering,* Stockholm, 1964: Militärpsykologiska institutet, report no. 31 (mimeo).

Etzioni, Amitai, *Winning Without War,* New York, 1965: Doubleday Anchor Books.
- *The Active Society,* New York, 1968: Free Press.

Feld, Maury D., "The Military Self-Image in a Technological Environment", in Janowitz, M. (ed.), *The New Military,* New York, 1964: Russell Sage Foundation.

Feuer, Lewis S., (ed.), *Marx and Engels: Basic Writings on Politics and Philosophy,* Garden City, 1959: Doubleday Anchor Books.

Finer, S.F., *The Man on Horseback,* London, 1962: Pall Mall Press.

Foertsch, Hermann. *The Art of Modern Warfare,* New York, 1940: Oscar Priest.

French, E.G., and R.B. Ernest, "The Relationship Between Authoritarianism and Acceptance of Military Ideology", *Journal of Personality,* vol. 24, 1955, pp. 181–191.

Frändén Olof, *Officersyrkets anseende bland gymnasister i Stockholm*, Stockholm, 1968: Militärpsykologiska institutet, B-report no. 8 (mimeo).
- "Notes on Mobility into and out of the Swedish Officer Corps", in van Doorn, J.A.A. (ed.), *Military Profession and Military Regines*, The Hague, 1969: Mouton.

Galbraith, John Kenneth, *The New Industrial State*, Boston, 1967: Houghton-Mifflin.
- *Att hålla Pentagon i schack*, Stockholm, 1970: Tema.
Garthoff, R.L., "The Military in Russia, 1861–1965", in van Doorn, J.A.A. (ed.), *Armed Forces and Society*, The Hague, 1968: Mouton.
Girardet, R., and J-P.H. Thomas, "Problèmes de Recrutement", in Girardet, R. (ed.), *La Crise Militaire Française, 1945–1962*, Paris, 1964: Librairie Armand Colin.
Goffman, Erving, "The Characteristics of Total Institutions", in Etzioni, A., *Complex Organizations – A Sociological Reader*, New York, 1964: Holt, Rinehart, and Winston.
Gould, Julius, and William L. Kolb, *Dictionary of the Social Sciences*, New York, 1964: Free Press.
Graczyk, Joseph, "Social Promotion in the Polish People's Army", in van Doorn, J.A.A. (ed.), *Military Profession and Military Regimes*, The Hague, 1969: Mouton.
Grape, Lennart, and Bengt-Christer Ysander, *Säkerhetspolitik och försvarsplanering*, Stockholm, 1967: Studieförbundet Näringsliv och Samhälle.
Greenwood, Ernest, "Attributes of a Profession", in Nosow, S., and W.H. Form, *Man, Work, and Society*, New York, 1962: Basic Books.
Göransson, Curt, "ÖB:s funktioner i fred och krig", *Fred och försvar under 60-talet*, Stockholm, 1961: Esselte.
Görlitz, Walter, *Der deutsche Generalstab*, Frankfurt a. M., 1950: Verlag det Frankfurter Hefte.

Hansen, Roy A., *Military Culture and Organizational Decline: A Study of the Chilean Army*. Unp. doct. diss., Department of sociology, University of California, Berkeley, 1967.
Hardin, Garrett A. (ed.), *Science, Conflict, and Society – Readings from Scientific American*, San Fransisco, 1970: W.H. Freeman & Co.
Hempel, C.G., "The Logic of Functional Analysis", in Brodbeck, May, (ed.), *Reader in the Philosophy of the Social Sciences*, New York, 1968: Macmillan.
Hobsbawn, E.J., *The Age of Revolution: Europe 1789–1848*, London, 1962: Weidenfeld & Nicholson.
Hollander, Edwin P., *Principles and Methods of Social Psychology*, New York, 1967: Oxford University Press.
Horowitz, I.L., *Three Worlds of Development*, New York, 1966: Oxford University Press.

— "The Military Elites", in Lipset, Seymour M., and A. Solari, (eds.), *Elites in Latin America,* London, 1967: Oxford University Press.

Hughes, Everett C., "Professions", in Lynn, K.S. (ed.), *The Professions in America,* Boston, 1965: Beacon Press.

Huntington, Samuel P., *The Soldier and the State,* Cambridge, Mass., 1957: Harvard University Press.

— "Power, Expertise, and the Military Profession ". in Lynn, K.S. (ed.), *The Professions in America,* Boston, 1965: Beacon Press.

Husén. Torsten, *Militärt och Civilt,* Stockholm, 1956: Norstedts.

Hyde, D.R., et al., "The American Medical Association", *Yale Law Journal,* vol. 63, 1954, no. 7.

de Imaz, J.L., *Los Que Mandan,* Buenos Aires, 1964: Editorial Universitaria de Buenos Aires.

Inkeles, Alex, and Peter H. Rossi, "National Comparisons of Occupational Prestige", *American Journal of Sociology,* vol. 61, 1955—56.

Jackson, John A., "The Irish Army and the Constabulary Concept", in van Doorn J.A.A. (ed.), *Armed Forces and Society,* The Hague, 1968: Mouton.

Janowitz, Morris, "Military Elites and the Study of War", *Journal of Conflict Resolution,* vol. I, 1957.

— *The Professional Soldier,* Glencoe, Ill., 1960: Free Press.

— *The Military in the Political Development of New Nations,* Chicago, Ill., 1964: University of Chicago Press.

Johnson, John J., *The Military and Society in Latin America,* Stanford, 1964: Stanford University Press.

Keller, Suzanne, *Beyond the Ruling Class,* New York, 1968: Random House.

Kennedy, Robert F., *Thirteen Days,* New York, 1969: Signet Books.

Kennett, Lee, *The French Armies in the Seven Years War,* Durham, N.C., 1967: Duke University Press.

Kjellberg, Francesco, "Some Cultural Aspects of the Military Profession", *European Journal of Sociology,* no. 2, 1965.

Kling, Merle, "Violence and Politics in Latin America", *The Sociological Review Monograph,* University of Keele, 1967.

Klockare, Sigurd, *Svenska revolutionen,* Stockholm, 1967: Prisma.

Kluckhohn, Clyde, *Mirror for Man,* New York and Toronto, 1949: McGraw-Hill.

Korpi, Walter, *Social Pressures and Attitudes in Military Training,* Stockholm 1964: Almqvist & Wiksell.

Krigsmaktens anda och ordning. Betänkande utgivet av ÖB, Stockholm 1947.

Lang, Kurt, *Sociology of the Military,* bibliography published by the Inter-University Seminar on Armed Forces and Society, 1969

Laski, Harold, *Democracy in Crisis,* Chapel Hill, 1933: University of North Carolina Press.

Lasswell, Harold D., and Abraham Kaplan, *Power and Society,* New Haven, 1950: Yale University Press.

Lemmel, C.F., "Rekryteringen till arméns aktiva officerskår ur intellektuellt kvalitativ synpunkt", in *Befälsordningen vid infanteriet,* 1953, SOU 1953:28.

Lerche, Charles O., "The Professional Officer and Foreign Policy", *Strategic Subjects Handbook,* U.S. Army Command and General Staff College, Fort Leavenworth, Kansas, September, 1967, R−1800−1, pp. L1−5 ff.

Lipset, Seymour M., and Mildred A. Schwarz, "The Politics of Professionals" in Vollmer, H.M., and Donald L. Mills, *Professionalization,* Englewood Cliffs, N. J., 1966: Prentice-Hall.

Lissak, Moshe, "Selected Literature on Revolutions and Coups d'etat in the Developing Nations", in Janowitz., M., *The New Military,* New York, 1964: Russell Sage Foundation.

− "Modernization and Role-Expansion of the Military in Developing Countries, *Comparative Studies in Society and History,* vol. IX, no. 3, April, 1967.

Lloyd George, David, *War Memories,* 1−VI, Boston, 1933−37.

Lorens, Konrad, *On Aggression,* New York, 1966: Harcourt, Brace, and World.

Lubove, Roy, *The Struggle for Social Security, 1900−1935,* Cambridge, Mass., 1968: Harvard University Press.

Lynn, K.S. (ed.), *The Professions in America,* Boston, 1965: Beacon Press.

McCarthy, Eugene, "The Power of the Pentagon", *Saturday Review,* December 21, 1968.

McClosky, Herbert, "Conservatism and Personality", *American Political Science Review,* vol. 52, 1958, pp. 27−45.

McKinley, R.D., *Professionalization, Politicization, and Civil-Military Relations,* paper, The Perceived Role of the Military, Social Science Symposium, Bandol, France, 1970.

MacRae, Duncan G., "Charisma", in Gould, Julius, and William L. Kolb, *Dictionary of the Social Sciences,* New York, 1964: Free Press.

Mannheim, Karl, *Ideology and Utopia,* New York, 1936: Harcourt, Brace, and World.

− *Man and Society in an Age of Reconstruction,* London, 1968.

Marsh, Raymond C., *Comparative Sociology,* New York, 1967: Harcourt, Brace, and World.

Maskerad front − Handbok i säkerhetstjänst. Stockholm, 1964: Försvarets Bok- och Blankettförråd.

Masland, John W., and Laurence I. Radway, *Soldiers and Scholars,* Princeton, New Jersey, 1957: Princeton University Press.

Merton, Robert K., *Social Theory and Social Structure,* Glencoe, Illinois, 1957: Free Press.

Miliband, Ralph, *Statsmakten i det kapitalistiska samhället,* Stockholm, 1970: Tema.
Millis, Walter, *Arms and Men,* New York, 1956: Mentor Books.
Mills, C. Wright, *The Power Elite,* New York, 1956: Oxford University Press.
Mosca, Gaetano, *The Ruling Class* (Elementi di Scienza Politica). English translation, edited and revised by Arthur Livingstone, New York an London, 1939: McGraw-Hill.

Nisbet, Robert A., *The Sociological Tradition,* New York, 1966: Basic Books.
Newman, James R., "Thermonuclear War", *Scientific American,* March, 1961.
Newton, K., "A Critique of the Pluralist Model", *Acta Sociologica,* vol. 12, 1969, no. 4.

Officersutbildningen inom armén m.m., Stockholm, 1946: SOU 1946:38. 1946:38.
Ortmark, Åke, *De okända makthavarna,* Stockholm 1970: Wahlström & Widstrand.
Otley, C.B., "Militarism and the Social Affiliations of the British Army Elite", in van Doorn, J.A.A. (ed.), *Armed Forces and Society,* The Hague, 1968: Mouton.
von Otter, Casten, et al., *Om hjälpstrukturer och vårdideologi i kapitalistiska samhällen,* unp. research paper, Department of Sociology, University of Uppsala, 1970 (mimeo.).

Parry, Geraint, *Political Elites,* London, 1969: George Allen & Unwin.
Parsons, Talcott, "Certain Primary Sources and Patterns of Aggression in the Social Structure of the Western World", in *ibid., Essays in Sociological Theory,* New York, 1964: Free Press.
Patterson, Robert P., Address before the Alumni Association of Columbia University, June 3, 1947", *Infantry Journal,* vol. 61, July, 1947.
Ponteil, F., *Napoleon 1er et l'organisation autoritaire de la France,* Paris, 1956: Librairie Armand Colin.
Power, Thomas S., *Design for Survival,* New York, 1965: Pocket Books, Inc.
Price, Don K., *The Scientific Estate,* New York, 1965: Oxford University Press.

Rapoport, Anatol, *Fights, Games, and Debates,* Ann Arbor, 1960: University of Michigan Press.
– "Introduction", in *Clausewitz on War,* London, 1968: Penguin.
Rehnberg, Mats, *Vad skall vi göra med de blanka gevär,* Stockholm, 1967: Nordiska Museet.
Reiss, Albert J., Jr., "Occupational Mobility of Professional Workers", *American Sociological Review,* vol. 20, 1955, pp. 693–700.

Richardson, Lewis F., "Generalized Foreign Policy", *British Journal of Sociology Monographs Supplement*, 23, 1939.

Riesman, David. *The Lonely Crowd*, New York, 1953: Doubleday.

Riksdagsmannavalen åren 1961–64:II, Stockholm, 1965, Statistiska Centralbyrån.

Roazen, Paul, *Freud – Political and Social Thought*, New York and Toronto, 1968: Alfred A. Knopf.

Roberts, Adam (ed.), *Civilmotståndets strategi*, Stockholm 1969: Aldus.

Rosenberg, Morris, *Occupations and Values*, Glencoe, Illinois, 1957: Free Press.

Rudé, George, *Det revolutionära Europa 1783–1815*, Stockholm, 1967: Aldus.

Russett, Bruce M., *World Handbook of Political and Social Indicators*, New Haven, Mass., 1964: Yale University Press.

Scott, Samuel F., *The French Revolution and the Professionalization of the French Officer Corps, 1789–1793*, paper, 7th World Congress of Sociology, Varna, Bulgaria, 1970.

Sherif, Muzafer, and Carolyn W. Sherif, *Groups in Harmony and Tension*, New York, 1966: Octagon Books.

Shoup, David M., "The New American Militarism", *The Atlantic*, April, 1969.

Smith, H.E., "What Is the Military Mind?", *U.S. Naval Institute Proceedings*, vol. 79, May, 1953.

Speier, Hans, *Social Order and the Risks of War*, New York, 1952: George W. Stewart.

Spindler, G.D., "The Military – A Systematic Analysis", *Social Forces*, vol. 27, 1958, pp. 83–88.

Stinchcombe, Arthur L., *Constructing Social Theories*, New York, 1968: Harcourt, Brace, and World.

Stouffer, S.A., *The American Soldier*, I, Princeton, 1949: Princeton University Press.

Stupak, Ronald J., "The Military's Ideological Challenge to Civilian Authority in Post-World War II France", *Orbis*, Summer, 1968, pp. 582–604.

Swomley, John M., Jr., *The Military Establishment*, Boston, 1964: Beacon Press.

Thoenes, Piet, *The Elite in the Welfare State*, New York, 1966: Free Press.

Tjänstereglemente för krigsmakten (TjrK), 1960: Försvarets Kommandoexpedition (Supplement, 1965).

Törnqvist, Knut, *Försvarsvilja och bedömning av krigsrisker*, Stockholm, 1966: Beredskapsnämnden för psykologiskt försvar (mimeo.).

Uppträdande utom tjänsten. Några råd och anvisningar för Kungl. Krigsskolans kadetter och arméns yngre officerare, Stockholm, 1963: Kungl. Krigsskolan (mimeo.).

Vagts, Alfred, *A History of Militarism,* New York, 1959, Hollis & Carter.

Veliz, Claudio (ed.), *The Politics of Conformity,* London, 1967: Oxford University Press.

Vollmer, H.M., and D.L. Mills (eds.), *Professionalization,* Englewood Cliffs, N.J., 1966: Prentice-Hall.

Waldman, Eric, *The Goose-Step is Verboten,* Glencoe, Ill., 1964: Free Press.

Whitehead, Alfred N., *The Adventure of Ideas,* New York, 1933: Macmillan.

Wiatr, J.J., "Military Professionalism and Transformations of Class Structure in Poland", in van Doorn, J.A.A. (ed.), *Armed Forces and Society,* The Hague, 1968: Mouton.

Wilensky, Harold L., "The Professionalization of Everyone?", *American Journal of Sociology,* vol. 70, 1964, pp. 137–158.

Williams, Robin M., *American Society,* New York, 1951: Alfred A. Knopf.

Wool, Harold, *The Military Specialist,* Baltimore, 1965: The Johns Hopkins Press.

Yarmolinsky, Adam, "The Problem of Momentum", in Chayes, Abram, and Jerome B. Wiesner (eds.), *ABM – An Evaluation of the Decision to Deploy an Anti-Ballistic Missile System,* New York, 1969: Signet Books.

Zhilin, P., "The Armed Forces of the Soviet State: Fifty Years of Experience in Military Construction", in van Doorn, J.A.A. (ed.), *Military Profession and Military Regimes,* The Hague, 1969: Mouton.

ÖB 65, Stockholm, 1965: Försvarets Bok- och Blankettförråd.

NAME INDEX

Abrahamsson, B. 48,55,67,90,97,101,103
Andreski, S. 24, 61, 105, 142

Bachrach, P. 113, 143
Bacon, Sir R. 86
Baldwin, H.W. 93
Baratz, M.S. 113, 143
Barber, B. 59, 65, 69
Barnes, J.W. 33, 34
Barnett, C. 29
Beaufre, A. 25, 26
Bendix, R. 43
Bengtsson, E.S. 84, 85
Benoit, E. 12
Bethe, H.A. 120
Bethmann-Hollweg 143
Blalock, H.M. 140, 145
Blau, P.M. 49, 55, 56
von Blomberg, W. 159
Boalt, G. 93
Borup-Nielsen, S. 52
Bottomore, T.B. 123
Boudet, J. 22, 109, 154
Bourcet 29
Bradley, O. 72, 82, 88
Brentano, L. 56
Burke, E. 104
Busquets Bragulat, J. 48, 50

Campbell, D.T. 94
Canton, D. 30
Carlsson, G. 96
Carr-Saunders, A.M. 14, 59
Chayes, A. 32, 110, 120, 155

von Clausewitz, C. 60
Coates, C.H. 59, 105, 160
Cook, F.J. 32, 115, 120
Craig, G.A. 22, 31, 41, 42, 125, 143, 154
Curcek, J. 48, 53

Dahl, R.A. 122, 140
Dahlgren, J.A.B. 32
Dahrendorf, R. 140
Dato 143
Delvigne 153
Demeter, K. 31, 41
Douglas, P. 155
Douglas, W.O. 93
von Doorn, J.A.A. 30, 44, 48, 51–53, 58, 60, 65, 162
Duncan, O.D. 49, 55, 56
Durkheim, E. 72

Eisenhower, D.D. 148
Encel, S. 48
Engels, F. 21, 24, 28
Ericson, S. H:son 62
Erikson, R. 73, 90, 98
Ernest, R.B. 94
Etzioni, A. 62, 86, 140

Feld, M.D. 120–121
Feuer, L.S. 24
Finer, S.E. 41, 143, 145
Foch, F. 109
Foertsch, H. 86
Form, W.H. 14, 59, 63
Fortescue, Sir J. 55
Frederick the Great 24, 31
Frederick William 23
Frederick William I 23
French, E.G. 94
von Fritsch, W. 159
Fråndén, O. 48, 52, 97–98

Galbraith, J.K. 20, 27, 32, 82, 118, 162
Garthoff, R.L. 58, 107, 108
Garwin, R.L. 120
Girardet, R. 48, 50
Goffman, E. 62
Goldwater, B.M. 110
Gorgas, J. 32
Gould, J. 59

Graczyk, J. 48, 54
Grape, L. 61
Greenwood, E.G. 14, 63, 66
Gribeauval 25
Guibert 29
Gustavus Adolphus 22
Göransson, C. 72, 82
Göring, H. 159
Görlitz, W. 33

Haig, D. 109
Halmos, P. 61
Hansen, R.E. 48, 50, 85
Hardin, G. 120
Hempel, C.G. 159
von Hindenburg, P. 42, 143
Hitler, A. 141, 159
Hobsbawm, E.J. 25
Hollander, E.P. 86
Horowitz, I.L. 36, 144
Hughes, E.C. 59, 60, 62, 66
Huntington, S.P. 59, 66, 75, 77, 78, 80, 85–90, 101, 108, 141,
    147–150, 152, 157–160
Husén, T. 93
Hyde, D.R. 64

de Imaz, J.L. 48
Inkeles, A. 96

Jackson, E. 28
Jackson, J.A. 48
Janowitz, M. 20, 28, 29, 35, 41, 43, 48, 59, 67, 68, 74, 78,
  81, 82, 93, 96, 101, 103, 110, 126, 142,
  147, 148–150, 152, 155, 162
Jefferson, T. 31
Johson, J.J. 30–32, 36, 44, 64, 80, 145
Jomini 60

Kahn, H. 60
Kaplan, A. 140
Kapp, W. 154
Keller, S. 13, 122, 123, 125
Kennedy, J.F. 148, 161
Kennedy, R.F. 161
Kennett, L. 26
Khan, A. 159
Kirk, R. 104

Kjellberg, F. 44, 48, 51, 52, 56–58, 60
Kling, M. 61
Klockare, S. 55
Kluckhohn, C. 86
Kolb, W.L. 59
Korpi, W. 78, 94
Kuczynski, R. 56

LaCierva 143
Lang, K. 20, 143
Laski, H. 72
Lasswell, H.D. 41, 140
LeMay, C. 142
Lemmel, C.F. 98
Lerche, C.O. 76, 77
Lipset, S.M. 43, 101
Lissak, Moshe 20, 36, 143
Lloyd George, D. 109
Lloyd, William 25
Lon Nol 126
Lorenz, K. 86
Louis XIV 23
Lubell, H. 12
Lubove, R. 123
Luckham, R. 160
Ludendorff, E. 36, 42, 143
Lundvall, B. 118
Lynn, K.S. 59, 60, 147

MacArthur, D. 63, 126, 145
McCarthy, E. 125
McClosky, H. 104–105
McCormack, T.H. 94
McKinlay, R.D. 13
McMurtrie, F.E. 86
McNamara, R.F. 120, 148
MacRae, D.G. 112
Mannheim, K. 72, 122
von Manteuffel, E. 22
Marx, K. 24
Masland, J.W. 13, 18, 126
Maury, M. 31
Merton, R.K. 72, 86
Miliband, R. 154
Millis, W. 24, 26–28, 31, 32, 33, 154
Mills, C.W. 31, 41, 68, 76, 77, 110, 154
Mills, D.L. 14, 59

Minić 153
Mirza, I. 159
von Maltke, H. 33, 35, 125
Moritz of Saxony 25
Mosca, G. 41–43, 57, 105

Napoleon Bonaparte 23–26, 29
Newman, J.R. 60
Newton, K. 123
Nisbet, R.C. 24
Nordenskiöld, C.H. 143
Nosow, S. 14, 59, 63
Nun, J. 36

Ortmark, Å. 61, 117
Otley, C.B. 48, 56
von Otter, C. 32

Parkenham 28
Parrott, R.P. 31
Parry, G. 41
Parsons, T. 80
Patterson, R.P. 71
Pellegrin, R.J. 60, 105, 160
Perón, J. 145
Power, T.S. 81, 142
Price, D.K. 82
Prieur 27, 153
Proxmire, W. 110, 155

Radway, L.F. 13, 18, 126
Rapoport, A. 75, 86
Rehnberg, M. 119
Reiss, A.D., Jr. 14
von Reyher 33
Rhee, Syngman 143
Richardson, L.F. 86
Riesman, D. 122
Roazen, P. 86
Roberts, A. 61
Rosenberg, M. 72, 73, 75, 78, 94, 105
Rossi, P.H. 96
Rossiter, C. 104
Rudé, G. 29
Russett, B.M. 142
Röhm, E. 141

Salisbury, Lord 89
Sarapata, A. 53
Schattschneider, E.E. 113
Schwarz, M.A. 101
Scott, S.F. 13, 21
von Seeckt 145
Sherif, C. 86
Sherif, M. 86
Sherman, W.T. 160
Shoup, D.M. 116
Smith, H.E. 72, 93
Song Yo Chan 143
Speier, H. 31, 36
Spindler, G.D. 62
Stinchcombe, A.L. 114, 120, 121
Stouffer, S. 29, 74, 93
Stupak, R.J. 125, 159
Swomley, J.M. 76, 77, 88

Taylor, T. 145
du Teil 29
Thoenes, P. 82
Thomas, J-P.H. 48, 50
Thouvenin 153
Torres 78
Tromp, H. 100
Truman, H.S. 87, 147
Twining, N.F. 142
Törnqvist, K. 92

Vagts, A. 21, 25, 28, 36, 41, 42, 55, 56, 60, 62,
  77, 78, 85, 88, 101, 105, 107, 109, 118,
  121, 159
Velasco 78
Veliz, C. 36
Vollmer, H.M. 14, 59
Waldman, E. 41, 48, 51, 78, 82, 83, 103
Whitehead, A.N. 62, 65
Whitney, E. 27
Wiatr, J.J. 53, 54
Wiesner, J.B. 32, 110, 120, 155
Wilensky, H.L. 65
Wilkes, C. 31
Williams, R.M. 114
Wilson, P.A. 14, 59

Yarmolinsky, A. 32, 110
York, H.F. 120
Ysander, B-C. 61

Zhilin, P. 58, 107, 108

SUBJECT INDEX

Aeroplanes 109–110
   production during World War I 109
Affective neutrality 109
Alarmism 77–78, 87–92, 99, 118, 160
  and adaptation to job 91
  and age 92
  and war scares in England 88
  civilian data 92
  covariation with level of professionalization 90
Army, mass 24, 26, 37, 39, 152
Army, standing 21, 41
Ascription 40–43, 56, 58
Ascriptive determinism 41, 43, 57
Authoritarianism 77–79, 93–95, 99
   and organizational form 95

Cavalry 28
Centralized state power 23–24
Civic action 36
Civilian control 66, 70, 147–150, 157–163
  and expectancy 145
  importance of professionalization for 17
  "objective" 157–163
  "subjective" 157–163
  through legislative supervision 161–162
Client
  concept of 64–66
  the state as 66, 69
Conflict research 61, 122
Conscription 22, 24, 152
Coups d'etat 18, 41–42, 57, 144, 154
Crystallization of attitudes 59, 76, 98, 112

Division
   composition of 25
   principle 25–26, 152

Expectancy 145, 149
   and civilian control 145, 149
   as component in mobilization 145

Fascism 107
Fundamentalism 80–81

Homogenization 72–75, 99
   and motivation 73
   and professional socialization 75
   and screening 73–74
   and selection 74

Industrialization
   and recruitment from farming population 55, 58
   and weapons production 26–27, 37, 39, 152

Levée en masse 22
Logistics, definition of 32–33
   as utilized by the German General Staff 33, 35
   development of 21–22, 23, 32–35
   functional importance of 39
   services in the US Army 34

Military academies 30, 50, 152
Military education 29–32, 38, 40, 59, 62, 151–152
Military engineers 29, 36, 153
Military establishment
   definition of 15
   formal organization 153
   similarity with civilian society 153–154
Military-industrial complex 20, 31–32, 65–66, 110, 123, 153–154
Military intervention 12, 17–18, 42, 57, 143, 154
Military manager 35–36, 40
Military mind 38, 71–79, 98–100
   as against civilian 71
   components of 76–78
Mobilization 144–147, 149
   as function of professionalization 145

Nationalism 62, 77, 78, 80–84, 99, 145, 153, 160
Non-decisionmaking 113
Normative influence 20, 112–128, 151, 155

defined 112
determinants of 115–120

Oaths of allegiance 63–64
Objective 114–115, 140, 145–147, 150, 156
   as component in mobilization 145
   defined 114–115
   in domestic policy 146–147
   in foreign policy 146
Occupational inheritance 49
   and special provisions for sons of officers 55–56, 58
Organization
   and expertise 152
   and institutionalization of values 113, 120–122, 127
   and profession, fusion of 65
   and professional values 78
Organizational differentiation 23, 32, 37, 39, 152

Pessimistic beliefs on human nature 77–78, 85–86, 99
Pluralism, critique of 122–124, 162
Political conservatism 42, 55, 78, 99, 101–111, 160
   and association with industry and business 108–110, 111
   and association with ruling elites 106–108, 111
   and peronality variables 104–106
   data on 102–103
   military as against other groups 101
Political role-expansion 18, 38, 57, 126, 155
Politicians and military issues 18, 161
Power, political 20, 112, 140–150, 151, 155
   defined 140
   different interpretations by Huntington and Janowitz 147–150
Power, centralized 23–24, 37, 39
Power of expertise 82
Power as against force 143
Prestige 53, 96–98
   and intellectual recruitment 97–99
   French data on 98
   Swedish data on 97
   U.S. data on 96
Profession, definition of 13–16
   and international communication 62
   and occupational inheritance 55
   and organization 65
Professional corporateness 15, 38, 43, 59, 68–69
   and control by peers 65
Professional education 62
Professional ethics 15, 59, 63–64, 69

Professional goals 78–79
Professional mind, definition of 74–75
Professional theory 15, 43, 59, 60–63, 69
Professionalization
   and ascriptive recruitment 56
   and autonomy 56, 65, 122–124
   and conservatism 107
   and mobilization of resources 145
   and politics 19
   and political intervention 37
   as allegedly increasing civilian control 157–163
   as historical process (professionalization$_1$) 16–17, 19, 21–39, 151–155, 157
   as homogenizing process (professionalization$_2$) 16–17, 19, 59–70, 72–75, 151, 155–156, 157
Public relations 115–116, 153

Resources 140–143, 148
   patterns of utilization 142–143
   quantitative aspects 142, 150
   structural aspects 141–142, 150

Standardization of weapons production 27, 152–153
Social recruitment 40–58, 152
   and class identifications 42
   and noble origins 21, 31, 37, 40, 41, 43, 56
   contemporary data on 46–47
   elitist 49–50, 57
   middle-class dominance in 44
   mixed 51–52, 57
   prospects 54–57
   research on 41–43
   rural 55, 58
   upper-class 50–51, 57
   working-class 52–53, 55, 57
Strategy and politics 13, 18, 124–127, 155

Tactics
   artillery 29, 153
   eighteenth-century 27
   in mass action 28

Values
   as component in mobilization 145
   definition of 114–115
   institutionalization of 112–114, 120–124, 127, 155–156
Veterans, U.S. 116